PERIODIC ELLIPTIC PARTIAL DIFFERENTIAL OPERATORS

The study of periodic partial differential equations (PDEs) has experienced significant growth in recent decades, driven by emerging applications in fields such as photonic crystals, metamaterials, fluid dynamics, carbon nanostructures, and topological insulators.

This book provides a uniquely comprehensive overview for mathematicians, physicists, and material scientists engaged in the analysis and construction of periodic media. It describes all the mathematical objects, tools, problems, and techniques involved. Topics covered are central for areas such as spectral theory of PDEs, homogenization, condensed matter physics, and optics. Although it is not a textbook, some basic proofs, background material, and references to an extensive bibliography providing pointers to the wider literature are included to allow graduate students to access the content.

Peter Kuchment is the Arthur George and Mary Emolene Owen Chair and Distinguished Professor at Texas A&M University. His research interests include PDEs, mathematical physics, geometric analysis, inverse problems, tomography and medical imaging, material science, and math education. He is the author of the books *Floquet Theory for Partial Differential Operators* (1993), *Introduction to Quantum Graphs* (2013), *The Radon Transform and Medical Imaging* (2014), and *Liouville-Riemann-Roch Theorems on Abelian Coverings* (2021).

CAMBRIDGE STUDIES IN ADVANCED MATHEMATICS

All the titles listed below can be obtained from good booksellers or from Cambridge University Press. For a complete series listing, visit www.cambridge.org/mathematics.

Already Published
183 A. Yekutieli *Derived Categories*
184 C. Demeter *Fourier Restriction, Decoupling and Applications*
185 D. Barnes & C. Roitzheim *Foundations of Stable Homotopy Theory*
186 V. Vasyunin & A. Volberg *The Bellman Function Technique in Harmonic Analysis*
187 M. Geck & G. Malle *The Character Theory of Finite Groups of Lie Type*
188 B. Richter *Category Theory for Homotopy Theory*
189 R. Willett & G. Yu *Higher Index Theory*
190 A. Bobrowski *Generators of Markov Chains*
191 D. Cao, S. Peng & S. Yan *Singularly Perturbed Methods for Nonlinear Elliptic Problems*
192 E. Kowalski *An Introduction to Probabilistic Number Theory*
193 V. Gorin *Lectures on Random Lozenge Tilings*
194 E. Riehl & D. Verity *Elements of ∞-Category Theory*
195 H. Krause *Homological Theory of Representations*
196 F. Durand & D. Perrin *Dimension Groups and Dynamical Systems*
197 A. Sheffer *Polynomial Methods and Incidence Theory*
198 T. Dobson, A. Malnič & D. Marušič *Symmetry in Graphs*
199 K. S. Kedlaya *p-adic Differential Equations*
200 R. L. Frank, A. Laptev & T. Weidl *Schrödinger Operators:Eigenvalues and Lieb–Thirring Inequalities*
201 J. van Neerven *Functional Analysis*
202 A. Schmeding *An Introduction to Infinite-Dimensional Differential Geometry*
203 F. Cabello Sánchez & J. M. F. Castillo *Homological Methods in Banach Space Theory*
204 G. P. Paternain, M. Salo & G. Uhlmann *Geometric Inverse Problems*
205 V. Platonov, A. Rapinchuk & I. Rapinchuk *Algebraic Groups and Number Theory, I* (*2nd Edition*)
206 D. Huybrechts *The Geometry of Cubic Hypersurfaces*
207 F. Maggi *Optimal Mass Transport on Euclidean Spaces*
208 R. P. Stanley *Enumerative Combinatorics, II (2nd Edition)*
209 M. Kawakita *Complex Algebraic Threefolds*
210 D. Anderson & W. Fulton *Equivariant Cohomology in Algebraic Geometry*
211 G. Pineda Villavicencio *Polytopes and Graphs*
212 R. Pemantle, M. C. Wilson & S. Melczer *Analytic Combinatorics in Several Variables (2nd Edition)*
213 A. Yadin *Harmonic Functions and Random Walks on Groups*
214 Y. Kawamata *Algebraic Varieties: Minimal Models and Finite Generation*
215 J. Gillespie *Abelian Model Category Theory*
216 L. Anderson *Oriented Matroids*
217 Y. Motohashi *Essays in Classical Number Theory*
218 H. L. Montgomery & R. C. Vaughan *Multiplicative Number Theory II*
219 E. Rijke *Introduction to Homotopy Type Theory*
220 N. Berestycki & E. Powell *Gaussian Free Field and Liouville Quantum Gravity*
221 J. Jost, R. Mulas & D. Zhang *Spectra of Discrete Structures*
222 D. Mubayi & J. A. Verstraete *Extremal Graph and Hypergraph Theory*
223 H. Sati & U. Schreiber *Equivariant Principal ∞-Bundles*

Periodic Elliptic Partial Differential Operators

The Main Structures

PETER KUCHMENT
Texas A & M University

CAMBRIDGE
UNIVERSITY PRESS

CAMBRIDGE
UNIVERSITY PRESS

Shaftesbury Road, Cambridge CB2 8EA, United Kingdom

One Liberty Plaza, 20th Floor, New York, NY 10006, USA

477 Williamstown Road, Port Melbourne, VIC 3207, Australia

314–321, 3rd Floor, Plot 3, Splendor Forum, Jasola District Centre,
New Delhi – 110025, India

Cambridge University Press is part of Cambridge University Press & Assessment,
a department of the University of Cambridge.

We share the University's mission to contribute to society through the pursuit of
education, learning and research at the highest international levels of excellence.

www.cambridge.org
Information on this title: www.cambridge.org/9781009282420
DOI: 10.1017/9781009282451

When citing this work, please include a reference to the DOI 10.1017/9781009282451

First published 2026

A catalogue record for this publication is available from the British Library

A Cataloging-in-Publication data record for this book is available from the Library of Congress

ISBN 978-1-009-28242-0 Hardback

To my family

Contents

Preface

Linear partial differential equations (PDEs) with periodic coefficients have been studied intensively since the early development of the quantum theory of solids. Today, they arise in many other scientific fields, including optics, waveguide theory, materials science, and fluid dynamics. Several monographs have been devoted wholly or in part to this subject. For example, see the books and surveys by Brown et al. (2013); Eastham (1973); Helffer and Sjöstrand (1989, 1990); Hess (1987); Karpeshina (1997); Kuchment (1982, 1993, 2016); Reed and Simon (1972); Rozenblyum et al. (1989, Section 17); Sjöstrand (1991); Skriganov (1985a); Titchmarsh (1958); and Veliev (2024). I should also mention monographs devoted to homogenization theory, which is closely related to the topic of this text, such as Bakhvalov and Panasenko (1989); Bensoussan et al. (1978); Conca et al. (1995); Jikov et al. (1994); Milton (2002, 2016); and Sánchez-Palencia (1980).

Why, then, another book on this subject?

First, many of these sources are outdated and do not cover recent substantial developments. Second, these monographs typically focus on a topic or, at most, a limited range of topics and do not represent a broad view of the field. I hope to provide a coherent, contemporary overview of this active area of research, as well as (in the planned sequel) an account of some of its most important applications and generalizations. My goal is to make the book accessible to a wide range of mathematicians and scientists interested in periodic PDEs. However, this is not a textbook; Chapter 1 provides only a brief overview of some classical 1D results. In addition, extensive references are provided for more detailed expositions of these topics.

This first volume of the planned monograph deals only with periodic elliptic operators, focusing on the development of the principal tools, approaches, and basic results. Other types of equations (parabolic, hyperbolic, evolution Schrödinger, etc.), as well as various applications and properties of solutions, are planned for treatment in the sequel (Kuchment, 2027).

Even with this restriction, some sections, or even chapters, provide only a brief overview of a topic and references, as a detailed exposition would warrant a separate large survey or a book. Examples are inverse problems, impurity spectra arising due to perturbations, "metallic" Wannier functions, or equations on abelian coverings.

A brief discussion of discrete periodic operators was initially planned, but to do justice to the rapid development of this area over the last decade would require a separate monograph.

A highly active and important area for research, both technically challenging and aesthetically pleasing, is devoted to the study of operators with limit-periodic, quasi-periodic, almost-periodic, and random coefficients. We do not consider such equations in this monograph. The reader can turn to many sources, for example, books and surveys by Aizenman and Warzel (2015); Bourgain (2005); Gesztesy et al. (2024); Carmona and Lacroix (1990); Chulaevsky (1989); Cycon et al. (1987); Damanik and Fillman (2022, 2024); Disertori et al. (2008); Marx and Jitomirskaya (2017); Pastur and Figotin (1992); Reed and Simon (1972); Shubin (1979) and Stollmann (2001) to find the relevant results and further references. This is far from being anywhere close to a more or less complete list of important publications in this area.

Acknowledgments

I am indebted to many colleagues and students for the benefit I have derived over the years from their works, discussions, and collaborations. The list of people who have influenced my work in this area is so long that I have decided not to test the readers' patience by listing all of them. Please know that I remember you all and am forever grateful. In particular, I owe a lot to my late teacher, S. Krein, who played a crucial role in my life and research.

The work on this monograph was partially supported by the US National Science Foundation (NSF) and the US-Israel Binational Science Foundation (BSF) grants, as well as by Texas A&M University. I express my gratitude to these institutions for their support. Part of the work was carried out during the author's several long-term visits to the Isaac Newton Institute (INI) for Mathematical Sciences in Cambridge. I express my gratitude to the INI for the support and wonderful work environment it provided.

This book would have never been written without the help and encouragement of the Cambridge University Press editorial staff, especially Mr. D. Tranah, Mr. Balaji Devadoss, and Ms. A. Serocka. I am also indebted to many colleagues and reviewers for their extremely helpful comments and suggestions, especially Gregory Berkolaiko, David Damanik, Jake Fillman, Leonid Friedlander, Evans Harrell, Yulia Karpeshina, Boris Mityagin, Stephan Shipman, and Selim Sukhtaev.

I am very grateful to my family for their patience and support.

Introduction

Elliptic partial differential equations (PDEs) with periodic coefficients, notably the stationary Schrödinger operator $-\Delta + V(x)$ with a periodic potential V,[1] have been intensively studied in physics and mathematics literature for close to a century, especially due to their central role in solid-state theory, as well as in other areas; see, for example, books and surveys by Ashcroft and Mermin (1976); Bethe and Sommerfeld (1933); Brillouin (1953); Brillouin and Parodi (1956); Kittel (1976); Reed and Simon (1972); Rozenblyum et al. (1989, §17); Ziman (1962, 1964). Despite that, some important questions remain unresolved, and new ones are popping up. The interest in periodic elliptic equations and systems (as well as in time-periodic parabolic, hyperbolic, or nonstationary Schrödinger equations and systems) has increased significantly in recent decades, due to new applications in areas such as photonic crystals and other metamaterials (Kuchment, 2001), fluid dynamics (Conca et al., 1995; Yudovich, 1989), carbon nanostructures (Katsnelson, 2012), inverse scattering method for solving integrable systems (Novikov, 1994; Novikov et al., 1984), and, more recently, topological insulators (Bernevig, 2013). While preparing introductory lectures at the Isaac Newton Institute (INI) in January 2015 (Kuchment, 2015), I realized that there was no comprehensive source devoted to the theory of periodic elliptic PDEs and their applications. There are several books and surveys completely or partially devoted to this topic (e.g., Allaire et al., 1998; Berezin and Shubin, 1991; Berkolaiko and Kuchment, 2013; Brown et al., 2013; Eastham, 1973; Gérard, 1990; Gieseker et al., 1993; Karpeshina, 1997; Kuchment 1982, 1993; Reed and Simon, 1972; Skriganov, 1985a; Titchmarsh, 1958; Veliev, 2024), but none of them provides a comprehensive and up-to-date account of the principal techniques (e.g., the analytic geometry of Bloch and Fermi varieties). Thus, the idea, which took years to mature,[2] emerged to expand the lectures into a book presenting the main techniques and results of the theory of periodic operators

[1] The potential is usually, but not always, assumed to be real.
[2] Meanwhile, many new results and extensions had been piling up.

arising in mathematical physics and related areas. These techniques apply not only to the standard model example of the Schrödinger operator with periodic electric potential $-\Delta + V(x)$ but also to a wide variety of elliptic periodic equations and systems, equations on graphs, the $\overline{\partial}$-operator, and other operators on abelian coverings of compact bases. Close attention must be paid to the analytic properties of Bloch and Fermi varieties, which significantly influence the behavior of such operators. The first stage in this expansion of the author's lectures at the INI was the survey by Kuchment (2016). This book is the second iteration, covering in more detail (and updating) the techniques described therein.

This monograph is *not intended to be a textbook*, so some background material commonly known to experts, especially concerning periodic ordinary differential equations (ODEs), is presented in a survey form, with extensive references. The goal of this volume is to provide an overview of the problems, techniques, and results of the theory of elliptic periodic PDEs. The author hopes that this monograph will be accessible to a broad range of mathematicians and researchers in other areas of science interested in periodic PDEs. Due to the diversity of problems and techniques applied, achieving this goal is somewhat of a challenge. The solution we chose was to defer some necessary background information and technical details of some proofs to the final Appendices A–E or refer to publications that contain complete proofs.

Another choice the author needed to make concerns the appropriate level of generality. Some results are known only for Schrödinger operators with electric potentials; many extend to any elliptic periodic operator, sometimes not even self-adjoint; and others may apply more broadly to systems in $\mathbb{R}^n$, overdetermined systems on manifolds, and even equations on graphs and quantum graphs (Berkolaiko et al., 2006). So, should one adopt the most general setting or, at the other extreme, restrict attention with Schrödinger operators only? It seems that both solutions would be wrong. For example, there have been many cases where the result was formulated for the Schrödinger operator only, while the proof clearly worked in great generality; some readers did not notice this generality of the technique. In addition, problems in optics, waveguide theory, and other areas require operators more general than the Schrödinger. On the other hand, making consideration as general as possible would make the text much harder to read and grasp. Thus, the solution chosen was somewhere in the middle. Namely, in most cases, in the main body of the text, a general scalar (usually second order[3]) elliptic operator in $\mathbb{R}^n$ is considered (unless the generality needs to be reduced further in some specific cases), although many results hold for systems or equations on manifolds. Hence, the readers should not think that the theorems stated in the book are provided in the

[3] As the reader will see, some important results might not hold for higher-order equations.

most general setting. Rather, one should consult the comment sections at the end of the chapters and to Appendices for further results and pointers to the literature, since in many cases more general and/or precise theorems are available.

Another simplification was adopted for clarity: the constructions are carried out for operators with "sufficiently smooth coefficients." In most cases, this assumption does not influence the main results and ideas of the proofs, but it sometimes eliminates substantial technical complications. This is not to say that studying the optimal conditions on coefficients is a worthless or simple exercise, since this would be definitely not true on both accounts. Moreover, in many applications, some coefficients are not merely not smooth but even discontinuous, so working with L^∞ (or even weaker) coefficients is often warranted. Efforts by many distinguished researchers have been made to optimize the conditions on the coefficients. However, we dare to say that, although obtaining results under precise conditions on the coefficients is in most cases a tough analytic work that requires true ingenuity, the main ideas of the theory presented in this book still stay the same. That is why we have decided to refer any reader interested in minimal requirements for the coefficients to the references and comments sections, where additional guidance to the literature is provided.

The reference list provided is extensive but surely far from comprehensive (which would be a daunting and nearly impossible task to achieve), and many references are cited through secondary sources, such as books and surveys. The author apologizes for this and for possible omissions.

One of the very first models with which to deal and the leading example in this book is the stationary Schrödinger operator,

$$H = -\Delta + V(x),$$

with a potential $V(x)$ that is periodic with respect to a lattice Γ in $\mathbb{R}^3$. Periodicity means that $V(x+\gamma) = V(x)$ for all $x \in \mathbb{R}^3$ and $\gamma \in \Gamma$. A major issue here (for the solid-state theory, see, e.g., Ashcroft and Mermin, 1976; Kittel, 1976; Ziman, 1964) is understanding the structure of the spectrum of the operator H in $L^2(\mathbb{R}^3)$. Although some spectral issues of importance to physics have been resolved mathematically rather quickly, others remained conjectural for a long time, widely believed but lacking rigorous proof. We can mention as an example the famous Bethe–Sommerfeld conjecture (Bethe and Sommerfeld, 1933) on the finiteness of the number of spectral gaps in dimensions higher than 1, a conjecture that has been settled only in the past two decades (Parnovski, 2008); see also Dahlberg and Trubowitz (1982); Helffer and Mohamed (1998); Karpeshina (1997); Parnovski and Sobolev (2010); Skriganov (1985a,b), as well as Veliev (1988, 2007, 2024).

Various issues are still unresolved, such as, for instance, the generic spectral-edge behavior of dispersion relations. This behavior is responsible, among other issues, for the existence of effective masses. Although the question of generic behavior has not yet been resolved, this certainly should not and does not stop physicists from using the notion of an effective mass. Many other studies, for instance, those related to homogenization, Anderson localization, or Liouville theorems, also push toward further investigation of the spectral structure of such periodic operators.

The introduction in 1987 of the idea of "photonic crystals" (or "photonic band-gap materials") (John, 1987; Yablonovitch, 1987) renewed interest in periodic elliptic equations. A photonic crystal is an optical analog of a semiconductor. It promises (and has already partially produced) a technological revolution in various areas of optics and electronics (e.g., Joannopoulos et al., 2008; Kuchment, 2001; Sakoda, 2001). It is conceived to be an artificially created, periodically structured dielectric medium. The equation governing propagation of electromagnetic waves in such a medium is the periodic Maxwell system. One is again interested in the spectrum, in particular the existence of spectral gaps for such operators. In many cases, the problems that arise are more difficult than those encountered for Schrödinger operators.

We could continue with such examples; some of them are discussed in more detail elsewhere in the book, in the surveys (Kuchment, 1982, 2016), and mostly in its planned sequel (Kuchment, 2027). This multitude of applications in the sciences explains the recent surge of interest in periodic PDEs. During the past decades, significant progress has been made in this area.

To keep the length of the book manageable, in several cases we had to provide only an overview and references to the literature, rather than a detailed discussion. Some of these topics will be addressed in more detail in the monograph (Kuchment, 2027) planned as a sequel.

If there is a technique that unifies most studies of periodic linear PDEs, it is the Floquet theory. Its role in this case is analogous to the one played by the Fourier transform for constant coefficient equations. It is essentially a Fourier transform, namely, the one with respect to the periodicity group Γ. Despite its importance, it is not a magic bullet. Even for constant coefficient equations in the whole space, one cannot say that applying the Fourier transform, and thus looking at the symbol of the operator, will easily resolve all problems. This is even less so in the periodic case. Nevertheless, any serious study must begin with the development of this technique, which is exactly what Chapters 1–3 are devoted to. Chapter 1 begins with a brief discussion of the classical Floquet theory for ODEs. It ends with the seemingly disappointing conclusion that the direct analog of this approach does not seem to work that well for PDEs. Fortunately, as the rest of the book shows, the theory can be successfully adapted and emerges even more elegant.

Chapter 2 addresses the main notions related to the multidimensional periodicity, such as lattices, fundamental domains, and dual lattices. Then in Chapter 3, the

Bloch and Floquet functions and the Floquet transform are introduced. The latter is the analog for the periodic case of the Fourier transform. Analogs of Plancherel and Paley–Wiener theorems are available. In analogy with the constant-coefficient case, these are invaluable tools throughout the book.

Chapter 4 starts with a study of the Bloch and Fermi varieties associated with a periodic elliptic operator. The Fermi variety (surface) is an analog of the characteristic variety of a constant coefficient operator (i.e., the set of zeros of its symbol). The Bloch variety additionally incorporates the spectral parameter. Thus, Fermi surfaces become sections of the Bloch variety, or in other words, the level sets of the multiple-valued function whose graph is the Bloch variety. This function (dispersion relation) plays a prominent role in the theory of periodic differential equations, as well as in applications to solid-state physics, photonic crystals, waveguide theory, and other areas. A sometimes useful incarnation of the Fermi surface, which we call Floquet surface, is also introduced.

As will be seen, a solid understanding of the geometry of these varieties (including their extensions to the complex domain) is critical to resolving many natural questions regarding the spectra and other properties of periodic operators. This is explained in Chapter 4.

A very significant part (both in terms of its importance and size) of the book is contained in Chapter 5, where problems of spectral theory are addressed. Section 5.1 establishes the basic fact about the essential self-adjointness of elliptic periodic operators. The reader then finds the results on the structure of the spectrum, the presence and number of spectral gaps, spectral asymptotics, Shnol' type theorem, Floquet–Bloch eigenfunctions, inverse problems, and more.

Chapter 6 discusses the appearance of impurity spectra arising from a localized perturbation. Wannier functions and the corresponding bases and frames are dealt with in Chapter 7.

Most of the material in this book could also be developed for periodic operators on abelian coverings of compact manifolds, instead of $\mathbb{R}^n$ covering the torus. However, this would have complicated the reading of the book. Consequently, a brief survey of the covering case and related references is deferred to Chapter 8. The case of non-abelian coverings is much trickier and significantly different from the abelian one, both in terms of techniques and results. We postpone its consideration for the sequel (Kuchment, 2027).

It is necessary to mention that, in some cases, the results are provided in the main text without any references or names. Additional references are then given in the comments section at the end of each chapter.

An attempt was made to provide a comprehensive reference list concerning periodic linear (hypo)elliptic equations. In some cases, references cite surveys and books rather than the original sources. Although the list of references is rather extensive, as is often the case, some important references might have been inadvertently omitted. The author apologizes in advance for such occurrences.

As the title indicates, the book is devoted to studying (linear) elliptic PDEs with periodic coefficients. The methods and results developed in this book mainly apply to equations of elliptic (and with some sacrifices to hypoelliptic, in particular, parabolic) types only.

There are many important generalizations and applications (e.g., time-periodic equations) of the topics discussed here in the elliptic case. For instance, time-periodic PDEs play an important role in fluid dynamics. Indeed, if one linearizes the Navier–Stokes equation near a time-periodic flow, one arrives at a parabolic equation with time-periodic coefficients. This is one of the reasons why a study of such equations was conceived (Yudovich, 1989). There is interest in studying time-periodic hyperbolic equations (e.g., wave equation) and nonstationary Schrödinger equation as well.

However, many methods and results that are available for elliptic equations do not readily (or at all) apply to these cases; thus, the author has postponed their discussion to the planned sequel. However, the author realized that it is hard to do justice in the same volume to these and many details of techniques that need to be employed. Thus, the task of a detailed and comprehensive discussion of such topics is postponed until the next monograph in preparation. However, even with this split, several topics considered in this book had to be addressed in brief survey sections and chapters, with references provided.

Tentative Contents of the Planned Sequel (Kuchment, 2027)

Periodic Differential operators
Generalizations and Applications

1. Properties of Solutions

- Floquet representations of solutions
- Liouville theorems
- Green's function asymptotics

2. Some Extensions and Generalizations

- Hypoelliptic equations
- Non-abelian periodicity

3. Time Periodicity

- Parabolic equations
- Hyperbolic equations
- Schrödinger equations

4. Applications

- Photonic crystals
- Waveguides
- Nanomaterials
- Topological insulators

1

Periodic Ordinary Differential Operators

In this chapter, we provide a brief overview of the main tool for studying periodic linear ordinary differential equations (ODEs) and systems of ODEs, commonly referred to as **Floquet** (or **Floquet–Lyapunov**) **theory** (Floquet, 1883; Lyapunov, 1896, 1899). Its basic principles are discussed in many textbooks on ODEs, for example, in Arnol′d (1988, 1978); Coddington and Carlson (1997); Coddington and Levinson (1955). Second-order periodic ODEs, such as the Hill equation, are of particular interest and are considered in many monographs and surveys, for example, Arscott (1964); Berezin and Shubin (1991, Section 2.8); Brown et al. (2013); Djakov and Mityagin (2006); Eastham (1973); Magnus and Winkler (1979); Reed and Simon (1972, Section XIII.16); Rozenblyum et al. (1989, Chapter 17); Simon (2011, Ch. 5); and Titchmarsh (1958, Ch. XXI) (see also Djakov and Mityagin, 2009 and references therein for the case of singular potentials). The volumes by Damanik and Fillman (2022, 2024) and Gesztesy et al. (2024) cover a much broader range of topics related to 1D Schrödinger operators, including, in particular, periodic operators. The remarkable though regrettably not widely known treatise by Yakubovich and Starzhinskii (1975) is also highly recommended. It offers a wealth of information concerning periodic ODEs and systems of ODEs with manifold applications, such as Hamiltonian and canonical systems, parametric resonance, and stability domains. In what follows, we highlight only a few basic aspects of Floquet theory, focusing on those that will be essential for the subsequent discussion.

1.1 Euler's Theorem for Constant Coefficient Equations

Consider a linear system of ODEs,

$$\frac{dx}{dt} = Ax, \qquad t \in \mathbb{R}, \quad x \in \mathbb{C}^n, \tag{1.1.1}$$

with a constant $n \times n$ matrix A and impose the initial condition $x(0) = x_0 \in \mathbb{C}^n$. Its solutions are easy to find:

$$x(t) = e^{tA} x_0.$$

This implies the following **Euler theorem** (Euler, 1743), found in any ODE textbook:

Theorem 1.1.1 *All solutions of* (1.1.1) *are linear combinations of the* ***exponential polynomial solutions****,*

$$x(t) = e^{\lambda t} \sum_{j \in \mathbb{Z}, j \geq 0} t^j p_j. \tag{1.1.2}$$

Here λ is an eigenvalue of A, $p_j \in \mathbb{C}^n$, and the sum is finite.

Notice that the nonzero powers j do arise when matrix A has nontrivial Jordan blocks. In the absence of such blocks (e.g., if A is self-adjoint or normal), the polynomial factors in (1.1.2) disappear and the solutions are linear combinations of exponential functions.

Floquet–Lyapunov theory, described in Section 1.2, provides an analog of this result for ODE systems with periodic coefficients.

1.2 Floquet–Lyapunov Theory

Let us consider now a linear system of ODEs with periodic coefficients:

$$\frac{dx}{dt} = A(t)x, \qquad t \in \mathbb{R}, \quad x \in \mathbb{C}^n, \tag{1.2.1}$$

where $A(t)$ is a 1-periodic $n \times n$ matrix-valued function.[1] We will not try to impose the weakest possible conditions on the entries of the matrix (Yakubovich and Starzhinskii, 1975, Ch. II) but assume that they are continuous.[2] Let $X(t)$ be the $n \times n$ matrix **fundamental solution** of (1.2.1). In more detail, let

$$e_j = (0, \ldots, 0, \underbrace{1}_{j\text{th entry}}, 0, \ldots, 0)$$

be the standard basis of $\mathbb{C}^n$ and $x_j(t)$ – the solution of (1.2.1) such that $x_j(0) = e_j$. Then $X(t)$ consists of the column vectors $x_j(t)$. One can think of $X(t)$ as the "propagation" operator (**propagator**) that "shifts by time t along the trajectories."

[1] All notions and results can be easily adjusted to any value of the period, for example, to the often preferrable 2π, by rescaling.

[2] The same strategy will be followed in the main body of the book. This is not to suggest that finding the optimal conditions on the coefficients of an equation is unimportant. On the contrary, it often becomes a really hard task. However, the main important structures arise even in the case of "the nicest possible" coefficients, and some of the questions that emerge are not easy to answer under the best of circumstances.

Namely, given an initial condition g at $t = 0$, the vector function $X(t)g$ provides the values of this solution at any time t.

To put it yet another way, $X(t)$ is the solution of the following initial-value problem for a matrix differential equation,

$$\frac{dX}{dt} = A(t)X(t), \qquad X(0) = I. \tag{1.2.2}$$

In the specific case where the value of t coincides with the period (that is, it is equal to 1 according to our assumption), we introduce the following notion.

Definition 1.2.1 The matrix

$$M := X(1) \tag{1.2.3}$$

of the shift along the trajectories of the system by the period is said to be the **monodromy matrix (or operator)** of the equation (1.2.1).

In other words, for a vector $g \in \mathbb{C}^n$, the vector Mg is the value at $t = 1$ of the solution $x(t)$ of (1.2.1) satisfying the initial condition $x(0) = g$.

Let now g be an eigenvector of M with an eigenvalue z. Due to the uniqueness theorem for the initial-value problem applied at $t = 1$, the eigenvalue is automatically nonzero and thus can be written as $z = e^{ik}$ for some $k \in \mathbb{C}$.

Definition 1.2.2 The eigenvalues $z := e^{ik}$ of the monodromy matrix M are called **Floquet multipliers** and numbers k are called **quasi-momenta** or **crystal momenta** (the latter names are derived from solid-state physics; Ashcroft and Mermin, 1976).

Remark 1.2.3 One notices that the value of the quasi-momentum k is defined only modulo $2\pi\mathbb{Z}$-shifts.[3]

One can see then that the solution of (1.2.1) with the initial value g at $t = 0$, where g is an eigenvector with eigenvalue z, has the following **quasi-periodicity** (also called **cyclicity** or **Floquet**) property:

$$u(t + \gamma) = z^{\gamma} u(t) = e^{ik\gamma} u(t)$$

for any $\gamma \in \mathbb{Z}$ and $t \in \mathbb{R}$. This solution then has the following **Bloch form**:[4]

$$u(t) = e^{ikt} p(t), \tag{1.2.4}$$

where the function $p(t) := e^{-ikt} u(t)$ is 1-periodic.

Since M is not necessarily normal, one has to consider also the possibility of the appearance of **generalized eigenvectors** and Jordan blocks. It is not hard to check (see, e.g., Iakubovich and Starzhinskii, 1975) that if g is a generalized eigenvector

[3] See the discussion of dual lattices and quasi-momenta in Chapter 2.

[4] The name is borrowed from physics. In mathematics, such solutions are called **Floquet or Floquet–Bloch solutions**, but we reserve this name for the more general type (1.2.5) of solutions.

corresponding to the eigenvalue $z = e^{ik}$, then the corresponding solution with the initial value g can be represented as follows:

$$u(t) = e^{ikt}\left(\sum_{j=0}^{N} t^j p_j(t)\right), \tag{1.2.5}$$

where the vector functions $p_j(t)$ are 1-periodic. We call solutions of this form **Floquet solutions**, so Bloch solutions correspond to the particular case when $N = 0$.

Remark 1.2.4 Compare with exponential polynomial solutions (1.1.2).

Reduction of the monodromy operator M to the Jordan form results in the following central result of Floquet theory (Theorem 1.2.5).

Theorem 1.2.5 (Floquet, 1883) *Every solution $u(t)$ of (1.2.1) is a linear combination of Floquet solutions, that is, can be represented as a finite sum*

$$u(t) = \sum_{l} e^{ik_l t}\left(\sum_{j=0}^{N_l} t^j p_{j,l}(t)\right), \tag{1.2.6}$$

where the vector functions $p_{j,l}(t)$ are 1-periodic.

Here N_l is the size of the Jordan block of the monodromy matrix corresponding to the eigenvalue e^{ik_j}. In the absence of Jordan blocks, only those terms with $j = 0$ are present, that is, one has an expansion into Bloch solutions.

This theorem is the basis of most studies of periodic linear ODEs (see, e.g., the treatise of Iakubovich and Starzhinskii, 1975). One can reformulate it in a somewhat stronger form as in Theorem 1.2.6.

Theorem 1.2.6 (Floquet's theorem) *There exist a 1-periodic matrix function $P(t)$ and a constant matrix C, such that the fundamental solution can be expressed as follows:*

$$X(t) = P(t)e^{Ct}. \tag{1.2.7}$$

Remark 1.2.7 Notice that

$$M = X(1) = e^{C}. \tag{1.2.8}$$

In fact, Theorem 1.2.8, even stronger than Theorem 1.2.6, holds:

Theorem 1.2.8 (Lyapunov reduction theorem) *There exists a periodic invertible matrix-function $P(t)$ such that the substitution $x(t) = P(t)y(t)$ reduces the system* (1.2.1) *to the one with a constant matrix C:*

$$\frac{dy}{dt} = Cy(t).$$

It is clear that this result, together with Euler's theorem 1.1.1, implies the Floquet theorem 1.2.5.

We introduce now the important notion of the Lyapunov exponent, which characterizes the maximal rate of exponential growth of solutions.

Definition 1.2.9 The **Lyapunov exponent** for a given ODE is the infimum of numbers r such that for any solution $u(t)$, there exists a constant C such that $|u(t)| \leq Ce^{rt}$.

It is clear that in the periodic case, the Lyapunov exponent is determined by the imaginary parts of the quasi-momentum:

$$\max_k(-\mathrm{Im}(k)). \tag{1.2.9}$$

1.3 Hill Operator

As it is well known from any undergraduate class on ODEs, that higher-order ODEs (in particular, those with periodic coefficients) can be rewritten as equivalent first-order systems. This, however, is not a very useful approach here, since it could obscure important specific features of the original equation that justify concentrating on some important cases. One finds an abundance of results of this kind in Iakubovich and Starzhinskii (1975). Probably, the most studied and important periodic ODE is **Hill's equation**:

$$-\frac{d}{dx}\left(p(x)\frac{du}{dx}\right) + q(x)u(x) = 0 \tag{1.3.1}$$

with (usually real) periodic coefficients[5] $p(x) > 0$ and $q(x)$. A lot of attention has been paid to this equation in many books and research papers, for example, in Eastham (1973); Iakubovich and Starzhinskii (1975); Magnus and Winkler (1979); Marchenko (2011); and Titchmarsh (1958). It has also played an important role in the inverse scattering technique for integrating nonlinear partial differential equations (PDEs) (McKean and van Moerbeke, 1975; McKean and Trubowitz, 1976; Novikov et al., 1984; Novikov, 1994).

We will concentrate now on a special case of the Hill operator, namely, the **Schrödinger operator with electric potential:** $V(x)$,

$$H = -\frac{d^2}{dx^2} + V(x), \tag{1.3.2}$$

which (and its higher-dimensional analogs) is of great relevance for the rest of this book.

[5] The reader probably has noticed that the independent variable is now denoted by x, rather than the time variable t. The reason is that from now on, the time all but disappears from our view.

We again impose the over-stringent condition of continuity of the periodic potential V (by default real, with some exceptions), although much weaker conditions often suffice (see, e.g., Djakov and Mityagin, 2006, Eastham, 1973, Iakubovich and Starzhinskii, 1975, Magnus and Winkler, 1979 and vol. 4 of Reed and Simon, 1972). Defined on the Sobolev space $H^2(\mathbb{R})$ as the domain, the operator is self-adjoint in $L_2(\mathbb{R})$ (see, e.g., Eastham, 1973). Indeed, if one defines the operator on the space $C_0^\infty(\mathbb{R})$, it will be essentially self-adjoint, with the only self-adjoint extension described above.

1.3.1 Monodromy, Hill Discriminant, and Such

We will often be interested in the spectral problem, given by

$$-\frac{d^2u}{dx^2} + V(x)u = \lambda u. \tag{1.3.3}$$

As Floquet theory suggests, let us consider the fundamental system of two solutions $\phi(x,\lambda), \psi(x,\lambda)$ (entire in λ) satisfying the following initial conditions:

$$\phi(0,\lambda) = \psi'_x(0,\lambda) = 1, \phi'_x(0,\lambda) = \psi(0,\lambda) = 0. \tag{1.3.4}$$

Lemma 1.3.1

(1) *The* ***monodromy matrix*** *is provided by the formula,*

$$M(\lambda) := \begin{pmatrix} \phi(1,\lambda) & \psi(1,\lambda) \\ \phi'_x(1,\lambda) & \psi'_x(1,\lambda) \end{pmatrix}. \tag{1.3.5}$$

(2)

$$\det M(\lambda) = 1. \tag{1.3.6}$$

Proof The first claim results from direct calculation. Concerning the second one, the determinant $\det M(\lambda)$ is the Wronskian of the fundamental system of solutions, introduced earlier, evaluated at $t = 1$. Since the Wronskian is time-independent and equal to 1 at $t = 0$, the statement (1.3.6) follows. □

Since we are interested in the eigenvalues of the monodromy matrix, the lemma suggests that all the pertinent information is contained in the trace of M:

Definition 1.3.2 The trace

$$\Delta(\lambda) := \operatorname{Tr} M(\lambda) = \phi(1,\lambda) + \psi'_x(1,\lambda), \tag{1.3.7}$$

is called the **Hill discriminant** (sometimes referred to as just **discriminant**), or the **Lyapunov function** of the operator (1.3.2).

Now, all Floquet multipliers z, being the eigenvalues of M, are found by solving the **secular equation**

$$z^2 - \Delta(\lambda)z + 1 = 0. \tag{1.3.8}$$

It provides Floquet multipliers and quasi-momenta for any given λ:

$$e^{ik} = z = \frac{1}{2}\left(\Delta(\lambda) \pm \sqrt{\Delta(\lambda)^2 - 4}\right). \tag{1.3.9}$$

An alternative form of the equation (1.3.8) that is sometimes convenient is as follows:

$$z + z^{-1} = \Delta(\lambda). \tag{1.3.10}$$

Theorem 1.3.3, which follows from (1.3.8), combines several basic statements found in usual sources (e.g., Eastham, 1973; Yakubovich and Starzhinskii, 1975). The potential is assumed to be real.

Theorem 1.3.3

(1) *If* $|\Delta(\lambda)| < 2$, *then*
   - *The Floquet multipliers are complex, distinct, of absolute value* 1, *and complex conjugate to each other.*
   - *In particular, the quasi-momenta are real and if* k *is a quasi-momentum, then* $-k$ *is also.*[6]
   - *All solutions of* $Hu = \lambda u$ *are bounded.*

(2) *If* $|\Delta(\lambda)| = 2$, *then there are two possible cases (depending on whether the graph of the discriminant intersects the levels of* ± 2 *or is tangential to them):*
   a $M(\lambda) = \pm I$, *and then*
      - *The two Floquet multipliers coincide and are equal to either* 1, *or* -1. *This is equivalent to having a periodic or antiperiodic eigenfunction.*
      - *The quasi-momentum is* $k = 0$ *or* $k = \pi \,(mod\, 2\pi)$. *(This is just a rephrasing of the previous statement.)*
      - *All solutions of* $Hu = \lambda u$ *are bounded.*
      - *The Lyapunov exponent is equal to zero.*

   b $M(\lambda)$ *has a Jordan block structure, and then*
      - *The quasi-momentum is either* $k = 0$, *or* $k = \pi \,(mod\, 2\pi)$.
      - *One of solutions of* $Hu = \lambda u$ *is bounded, and one grows linearly.*
      - *The Lyapunov exponent is equal to zero.*

(3) *If* $|\Delta(\lambda)| > 2$, *then*

[6] This observation also follows from the fact that, due to the potential being real and for real λ, the complex conjugate of a solution is a solution as well. In physics terms, this is a reflection of the **time reversibility** of the dynamical system with the Hill operator as the Hamiltonian.

- *The Floquet multipliers are real, distinct, and reciprocal to each other.*
- *Quasi-momenta are complex.*
- *The solutions of $Hu = \lambda u$ grow exponentially.*
- *The Lyapunov exponent is positive and equal to* $|\mathrm{Im} k|$.

To visualize the behavior of the discriminant, let us look first at the **free operator** (i.e., $V = 0$). One computes that in this case, the discriminant is given by

$$\Delta_0(\lambda) = 2\cos(\sqrt{\lambda})$$

(see its graph in Fig. 1.1), which is an entire function of exponential order $1/2$, that is,

$$|\Delta_0(\lambda)| \leq Ce^{C|\lambda|^{1/2}}$$

for some $C > 0$. One notices that the extrema of the free discriminant are located

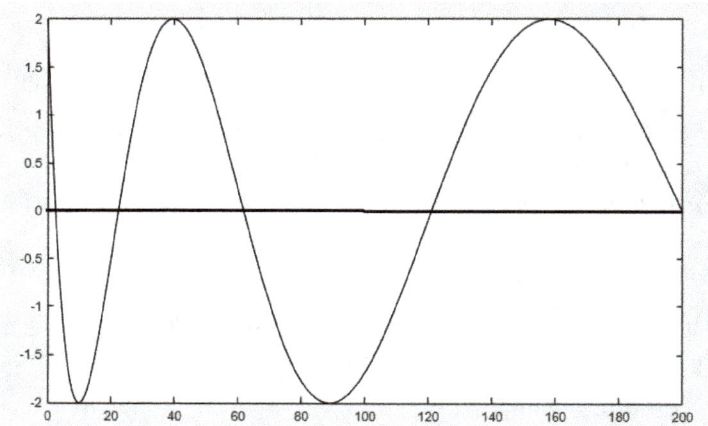

Figure 1.1 The graph of the free discriminant.

at the boundary of the horizontal strip $|\Delta| \leq 2$.

The properties of the discriminant, when the periodic potential is not equal to zero, have also been studied well (see, e.g., Eastham, 1973, Magnus and Winkler, 1979, Titchmarsh, 1958). As in the free case, it is an entire function of exponential order $1/2$:

$$|\Delta(\lambda)| \leq Ce^{C|\lambda|^{1/2}}.$$

Figure 1.2 gives an idea of its behavior along the real axis.

We observe that the extrema get shifted up or down (without any new ones being created). However, they cannot enter the interior of this strip (see detailed considerations, e.g., in Eastham, 1973). In particular, the function $\Delta(\lambda)$ is **monotonic** between

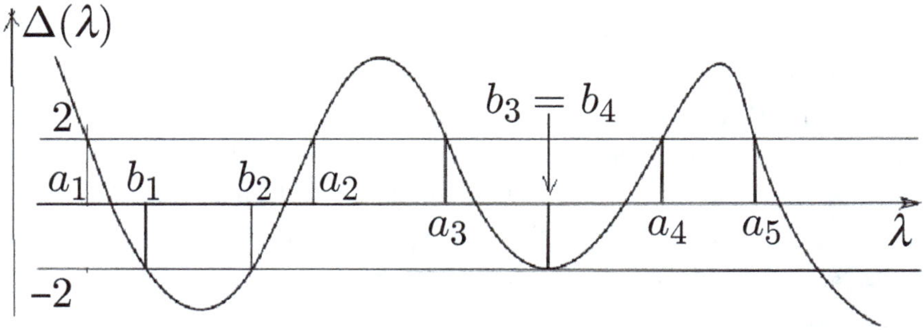

Figure 1.2 Discriminant of a periodic Hill equation.

its consecutive hits of the lines $\Delta = \pm 2$. A simple explanation of this monotonicity will be given a little bit later. As we will realize, absence of such monotonicity in higher dimensions leads to significant changes in spectral properties of periodic PDEs in comparison with the 1D case.

Definition 1.3.4 The closures of the segments where $|\Delta(\lambda)| < 2$ (the segments $[\lambda_{2n}, \mu_{2n}]$ and $[\mu_{2n+1}, \lambda_{2n+1}]$ in Fig. 1.3) are called the **stability zones** or **spectral bands**. The segments $(\lambda_{2n+1}, \lambda_{2n+2})$ and (μ_{2n}, μ_{2n+1}) between the bands are called **instability zones** or **spectral gaps**.

We will use the notation $g_n \geq 0$ for the length of the nth gap, $n = 1, 2, \ldots$.

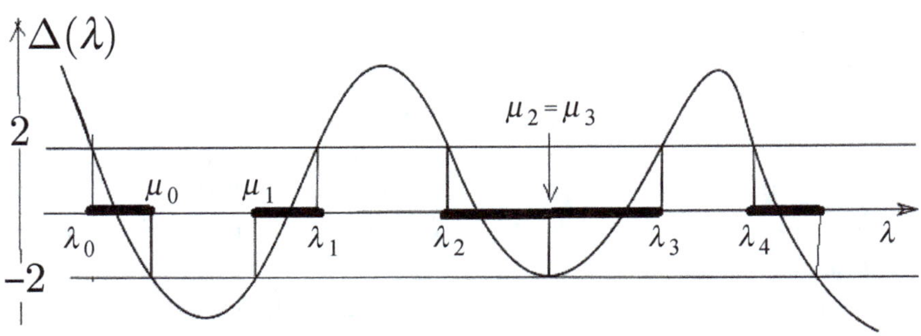

Figure 1.3 Stability zones (spectral bands), shown as thick lines, are separated by the instability zones (spectral gaps).

Remark 1.3.5 We conclude that the spectral bands are characterized by the quasi-momentum being real and the Lyapunov exponent being equal to zero. Correspondingly, the spectral gaps are characterized by the quasi-momentum being complex and the Lyapunov exponent positive.

1.3.2 Spectrum of the Hill Operator

Hold on! We have used the name "spectral band," before even discussing the spectrum of the Hill operator $H = -d^2/dx^2 + V(x)$ at all! No worries, it is not hard to show (and will be done later in higher dimensions) that the spectrum $\sigma(H)$ indeed coincides with the union of all spectral bands. In particular, the existence of bounded generalized solutions inside the spectral bands allows one to show that the bands belong to the spectrum. Indeed, cutting off these solutions at infinity one can provide approximate eigenfunctions. In fact, even in the PDE case, existence of a sub-exponentially growing solution indicates the spectrum (the so-called Shnol' Theorem 5.8.3 discussed later on). By contrast, one can show that inside the gaps, there is no spectrum.

Let us gather together the major features of the spectrum (the potential is assumed to be real).

Theorem 1.3.6

(1) *There exists a sequence of real numbers*

$$\lambda_0 < \mu_0 \le \mu_1 < \lambda_1 \le \lambda_2 < \cdots \mapsto \infty \tag{1.3.11}$$

(ends of the spectral bands) such that the spectrum $\sigma(H)$ coincides with the union of all spectral bands I_j

$$\sigma(H) = \bigcup_{j=0}^{\infty} I_j, \tag{1.3.12}$$

where

$$I_0 = [\lambda_0, \mu_0],\ I_1 = [\mu_1, \lambda_1],\ I_2 = [\lambda_2, \mu_2], \ldots . \tag{1.3.13}$$

(2) *The numbers λ_j (μ_j) are the eigenvalues of the operator restricted to the single period and equipped with periodic (respectively, antiperiodic) boundary conditions.*

(3) *The spectral bands have finite lengths and* **do not overlap** *but might touch (see Lemma 1.3.10).*

(4) ***Generically*** *(in the sense of the Baire category) with respect to the C^∞ potential,* ***all gaps are present (open)****, that is, there are no equalities in* (1.3.11), *so the bands do not touch* (Colin de Verdière, 1991; Simon, 1976).

(5) ***Finite gap (finite-zone) potentials****, that is, those that lead to only finitely many open gaps are special and can be all described* (Dubrovin et al., 2004, 2001; Novikov, 1994; Novikov et al., 1984) *At the same time, they form a dense set in the space of all C^∞-potentials* (Colin de Verdière and Kappeler, 1989; Marchenko and Ostrovsky, 1980, 1988).

(6) *There are no open gaps if and only if the potential is constant: the celebrated* ***Borg uniqueness theorem*** (Borg, 1946).

(7) *The rate of decay of the* ***gaps' sizes*** g_n *when* $n \to \infty$ *determines smoothness of the potential* (Djakov and Mityagin, 2002, 2006, 2009; Marchenko and Ostrovsky, 1975). *For instance,* C^∞ *potentials correspond to the gap size decaying faster than any power of 1/*n. *See, for example,* Djakov and Mityagin (2009, pp. 96–97) *for detailed history.*

(8) *There exist* ***isospectral*** *potentials. The sets of isospectral finite-zone potentials form tori* (McKean and van Moerbeke, 1975; McKean and Trubowitz, 1976; Novikov, 1994).

1.3.3 Dispersion Relation

Here we introduce, in the particular case of Hill operators, the crucial notion of the **dispersion relation**. It is rarely discussed in the literature devoted to periodic ODEs (with few exceptions, see the papers by Firsova (1975, 2000) concerning the **Riemann surface of quasi-momentum**, which has an immediate relation to the dispersion relation), but it is crucial for solid-state physics and for studying periodic PDEs in general.

Simply put, the dispersion relation is the spectral parameter λ as a (multiple-valued) function of the quasi-momentum (crystal momentum) k.

Definition 1.3.7 The **dispersion relation** (or the **Bloch variety** B_H) of the Hill operator H is the subset of $\mathbb{R}_k \times \mathbb{R}_\lambda$ defined as follows:

$$B_H := \{(k, \lambda) \in \mathbb{R}^2 \mid Hu = \lambda u \text{ has a non-trivial Floquet–Bloch solution } u(x) = e^{ikx}p(x) \text{ with the quasi-momentum } k, \text{ and periodic } p(x) \tag{1.3.14}$$

The **complex dispersion relation** (or **complex Bloch variety**) $B_{H,\mathbb{C}}$ is defined analogously, now allowing both k and λ to be complex.[7]

The following statements (Theorem 1.3.8) are not difficult to prove:

Theorem 1.3.8

(1) *In terms of the discriminant, the dispersion relation (both in the real and complex incarnations) can be described as the set of all (correspondingly real or complex) solutions* (k, λ) *of the equation*

$$\Delta(\lambda) - 2\cos k = 0. \tag{1.3.15}$$

[7] One might think that the complex values of the spectral parameter (energy) and quasi-momentum are somewhat "esoteric" or "nonphysical." This impression is incorrect. It will be clear from many future considerations how important is the ability to venture into the complex domain.

In other words, it is the graph of the multiple-valued function

$$\lambda = \Delta^{-1}(2\cos k). \tag{1.3.16}$$

(2) *The dispersion relation is 2π-periodic with respect to k.*
(3) *The real dispersion relation is even with respect to k. The complex one is invariant with respect to the mapping $(k, \lambda) \mapsto (-k, \overline{\lambda})$.*[8]
(4) *If the potential is real, the spectrum of the operator H coincides with the range of the multiple-valued function $\lambda(k)$, that is, with projection of the (real) dispersion relation onto the spectral λ-axis.*
(5) *If the branches of the multiple-valued function $k \in \mathbb{R} \mapsto \lambda \in \mathbb{R}$ are labeled in increasing order ($\lambda_1(k) \leq \lambda_2(k) \leq \cdots$), then the range of the jth branch is the jth spectral band of H.*

Formula (1.3.16) is less useful than (1.3.15), due to its multiple-valued (and branching) nature. Moreover, an important observation is that

$$F(\lambda, k) := \Delta(\lambda) - 2\cos k \tag{1.3.17}$$

is an entire function in $\mathbb{C}^2$ of exponential order 1 with respect to k and $1/2$ with respect to λ. Since it defines the dispersion relation as the set of all its zeros, one can make the following observation, the analog of which gains prominence in the PDE situation.

Proposition 1.3.9 *The dispersion relation of the Hill operator is a principal (i.e., of co-dimension one) analytic subset*[9] *in $\mathbb{C}^2$.*

Let us examine the free case ($V = 0$). Then (1.3.15) becomes $\cos\sqrt{\lambda} = \cos k$. Due to the periodicity with respect to k, it is convenient to draw this graph over the interval $[-\pi, \pi]$ only.[10] The graph of this part of the dispersion relation is shown in Fig. 1.4. This graph might seem somewhat complex, but this complexity is illusory. Indeed, this dispersion relation is just the union of infinitely many independent shifts of a parabola:

$$\lambda = (k + 2n\pi)^2, n \in \mathbb{Z}, \tag{1.3.18}$$

see Fig. 1.5.

Due to the evenness with respect to the quasi-momentum k (time-reversal symmetry), one can, without loosing any information, restrict attention to the **reduced Brillouin zone** $[0, \pi]$ only.

[8] Indeed, since the coefficients of the operator are real, the complex conjugate of a Floquet solution is also a Floquet solution. In terms of the non-stationary Schrödinger equation, this corresponds to the reversibility of the equation in time. There are important cases, though, for example, the presence of a magnetic potential, where this symmetry is broken.

[9] The additional information that the defining function F is an entire function of a finite exponential order sometimes becomes very significant.

[10] Later on, this interval will be called a **Brillouin zone**.

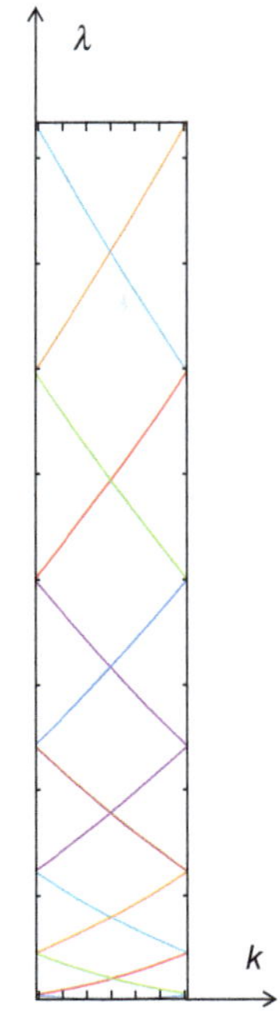

Figure 1.4 The dispersion relation of the free operator over $[-\pi, \pi]$.

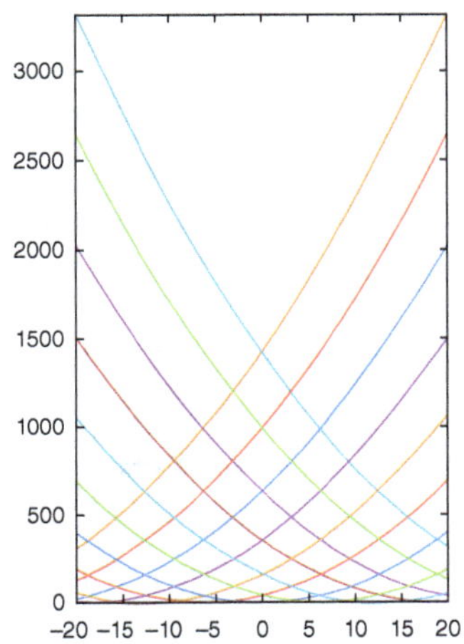

Figure 1.5 The dispersion relation over a segment of the whole quasi-momentum line of the free operator.

Turning on a small periodic potential $V(x)$ surely perturbs the dispersion relation, as shown in Fig. 1.6. You see that under this perturbation, some spectral gaps are opening. Also, it seems that, in spite of the perturbation, over $[0, \pi]$ each

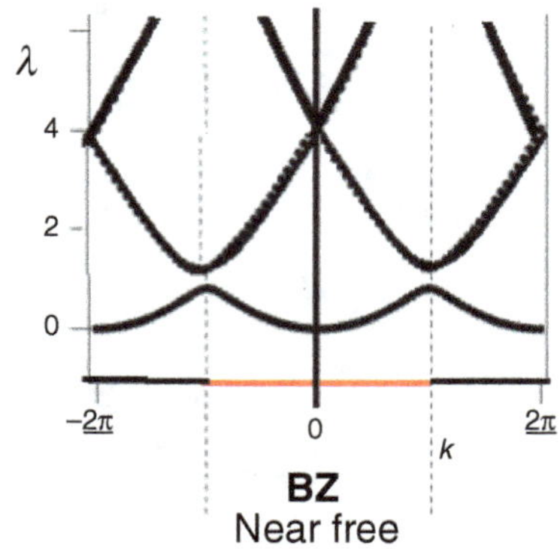

Figure 1.6 The dispersion relation in presence of a small periodic potential.

eigenvalue branch $\lambda_j(k)$ is still monotonic. This is indeed true, as the following result shows:

Lemma 1.3.10

(1) *Each branch $\lambda_j(k)$ of the dispersion relation is monotonic over $[0, \pi]$.*
(2) *In particular, the extrema of the band functions, and thus spectral band edges, occur only at $k = 0$ and $k = \pi$, that is, for solutions of periodic or antiperiodic problems for the Hill operator.*
(3) *The spectral bands of the Hill operator do not overlap (although they might touch, as the example of $V = 0$ shows).*
(4) *Each branch $\lambda_j(k)$ is continuous and piecewise analytic, losing its analyticity only at the collision points with other branches (= multiple eigenvalues).*

Proof Indeed, if the first claim were not true, taking into account the evenness with respect to k, one would have a situation as shown in Fig. 1.7. If we pick the value of λ at the level shown, we will find **four** values of k where this level is reached. This means that equation $Hu = \lambda u$ has four linearly independent solutions. Since the ODE is of second order, this is impossible.

This implies the second claim of the lemma.

The same argument shows the impossibility of band overlaps, as shown in Fig. 1.8. Finally, the last claimed property is the statement of the well-known result from Kato–Rellich perturbation theory (see, e.g., Theorem XII.8 in Reed and Simon, 1972, and references therein). □

Remark 1.3.11 The conclusions of the lemma were drawn from the fact that a second-order ODE cannot have more than two linearly independent solutions. One might wonder whether this is just an artifact of our proof, or something deeper. For

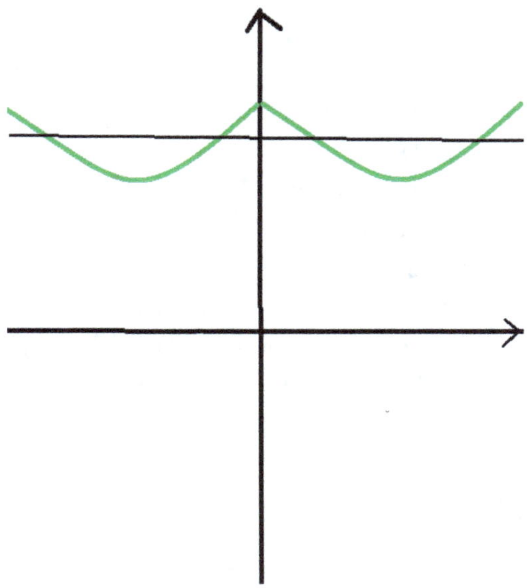

Figure 1.7 Non-monotonicity is impossible.

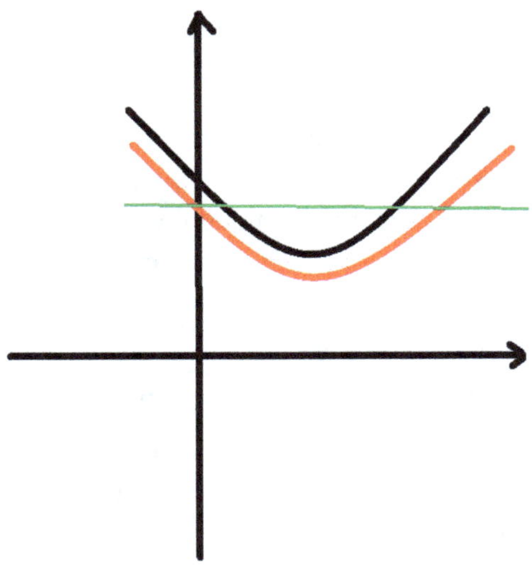

Figure 1.8 No band overlap.

instance, this argument does not apply to PDEs. Indeed, we will see that for PDEs, the first three claims are in general false.

We formulate now without proofs some additional spectral properties:

Theorem 1.3.12

(1) *The pure point and singular continuous spectra of the Hill operator are empty. Thus, the spectrum is absolutely continuous*[11] (Eastham, 1973; Kuchment, 1993; Reed and Simon, 1972).
(2) *The dispersion relation, as an analytic set, is irreducible (modulo the 2π shifts of the quasi-momentum). In other words, any open part of the dispersion relation (and its shifts by periods) uniquely determines the whole of it. To put it still differently, the smooth part of B_H is connected (*Kohn, 1959*; see also* Avron and Simon, 1978 *and* Rozenblyum et al., 1989, §17*).*
(3) *The Bloch variety (which is analytic, as we know) is generically not algebraic (e.g.,* McKean and van Moerbeke, 1975; McKean and Trubowitz, 1976*).*

1.3.4 1D Periodic Laplace–Beltrami Operators

Suppose that the real line is supplied with a periodically varying Riemannian metric, then one can consider the **Laplace–Beltrami operator** L with respect to this metric, which is a linear second-order differential operator with periodic coefficients. It reduces to

$$L := -\left(\frac{1}{\sqrt{a}}\frac{d}{dx}\right)^2 \tag{1.3.19}$$

with a positive function $a(x)$. A simple change of variables reduces it to $-\frac{d^2}{dy^2}$. Thus, its spectrum coincides with the positive half-axis and, unlike in the Schrödinger operator case, has no gaps. Interestingly enough, we will see in Chapter 5 (Section 5.4.4) that in higher dimensions, periodic Laplace–Beltrami operators can have multiple spectral gaps.

1.4 Some Quantitative Characteristics of the Spectrum

Some of the notions introduced here (e.g., the spectral function, or integrated density of states (IDS)) also apply to the multidimensional case, while some others (global quasi-momentum or rotation number) do not.

1.4.1 Global Quasi-momentum

Definition 1.4.1 The function $k(\lambda)$ in $\mathbb{R}$ is called the **global quasi-momentum** if it satisfies the following conditions:

(1) it is continuous and non-decreasing;
(2) it is zero below the spectrum;

[11] We will discus this in detail for the PDE case.

(3) it is constant in each spectral gap;
(4) it satisfies the equality

$$2\cos k(\lambda) = \Delta(\lambda) \tag{1.4.1}$$

on the spectrum.[12]

The behavior of the Hill discriminant $\Delta(\lambda)$ (see Figure 1.3) shows the existence and uniqueness of the global quasi-momentum.

One can establish the following lemma:

Lemma 1.4.2 *Assuming that the period is equal to* 1, *in the closure of the mth spectral gap (the semi-infinite one is the zeroth one) one has*

$$k(\lambda) = m\pi; \tag{1.4.2}$$

for a period of length a, one has $k(\lambda) = m\pi/a$.

One of the uses of the global quasi-momentum is to get a formula for the **spectral function** (i.e., the kernel of the spectral projector) for the Hill operator. Indeed, consider for any $\lambda \in \sigma(H)$ the normalized Bloch solution $\psi_\lambda(x)$ with quasi-momentum $k(\lambda)$ of the equation $Hu = \lambda u$. Then the solution $\overline{\psi_\lambda}$ has the quasi-momentum $-k(\lambda)$ (and in the interior of the spectral bands, the two are linearly independent).

Theorem 1.4.3 *The spectral function of the Hill operator is given by the formula*

$$e(\lambda; x, y) = \frac{1}{\pi}\int_{-\infty}^{\lambda} \mathrm{Re}[\psi_\mu(x)\overline{\psi_\mu(y)}]\, dk(\mu). \tag{1.4.3}$$

Due to the 1-periodicity, the spectral function satisfies the identity

$$e(\lambda; x+1, y+1) = e(\lambda; x, y). \tag{1.4.4}$$

Note that in this definition, the global quasi-momentum is defined on the real axis only. A different notion of a **global quasi-momentum** can be introduced by considering the quasi-momentum k as a double-valued function of the spectral parameter λ in the whole complex plane. This function has branch point singularities at the ends of the spectral zones (Avron and Simon, 1978; Kohn, 1959). One can describe the Riemann surface of this function as two copies of the complex plane, properly glued, with cuts along the bands of the spectrum (see Kohn, 1959, Firsova, 1975, and the survey Firsova, 2000, with more detailed references therein). This construction turns out to be very useful for the inverse scattering problem for the perturbed Hill operator, as is described in the above-referenced survey.

In higher-dimensional case, the dimensionality of the quasi-momentum k makes

[12] Compare with (1.3.15).

it inconvenient to consider it as a function of the spectral parameter λ. Thus, the dispersion relation $\lambda(k)$ takes its place.

1.4.2 The Rotation Number

Definition 1.4.4 Let ψ be a non-trivial real solution of $H\psi = \lambda\psi$. The rotation number $\omega(\lambda)$ is defined as follows:

$$\omega(\lambda) := -\lim_{x\to\infty} \frac{1}{2\pi x} \mathrm{Arg}(\psi(x) + i\psi'(x)), \tag{1.4.5}$$

where Arg is any branch of the argument continuous in x.

One can show (Johnson and Moser, 1982, 1983) that the rotation number exists and is independent of the choice of solution ψ. Moreover, it has properties eerily similar to the global quasi-momentum, namely:

Theorem 1.4.5 (Johnson and Moser, 1982, 1983) *The rotation number $\omega(\lambda)$, as a function of $\lambda \in \mathbb{R}$, has the following properties:*

(1) *it is continuous and non-decreasing;*
(2) *it is equal to zero below the spectrum and constant in each spectral gap.*

It stands to reason to expect a relation between the rotation number and the global quasi-momentum. And indeed,

Theorem 1.4.6

$$\omega(\lambda) = \frac{1}{2\pi} k(\lambda). \tag{1.4.6}$$

1.4.3 Density of States

Definition 1.4.7 The **Integrated Density of States (IDS)** (or the **spectral counting function**[13]) $N(\lambda)$ is defined as follows:

$$N(\lambda) := \int_0^1 e(\lambda; x, x) dx. \tag{1.4.7}$$

Due to (1.4.4), the function $e(\lambda; x, x)$ is 1-periodic in x, and thus $N(\lambda)$ is its average over the period.[14]

Theorem 1.4.8

[13] These names can be explained as follows. Let us restrict the operator to a finite segment $[0, L]$ and impose some fixed self-adjoint boundary conditions (e.g., periodic). Then the spectrum will be discrete and we can consider the usual eigenvalue counting function $N_L(\lambda)$. Then $N(\lambda) = \lim_{L\to\infty} L^{-1} N_L(\lambda)$.

[14] If the period is equal to a, rather than 1, one needs to change the formula (1.4.7) to $N(\lambda) := a^{-1} \int_0^a e(\lambda; x, x) dx$.

(1) *$N(\lambda)$ is non-decreasing;*
(2) *$N(\lambda) = 0$ below the spectrum and constant inside each spectral gap;*
(3) *the spectrum $\sigma(H)$ coincides with the set of growth points of $N(\lambda)$:*

$$\sigma(H) = \{\lambda \mid N(\lambda + \epsilon) - N(\lambda - \epsilon) > 0 \text{ for any } \epsilon > 0\} \tag{1.4.8}$$

(4)

$$N(\lambda) = \frac{1}{\pi} k(\lambda). \tag{1.4.9}$$

Remark 1.4.9 In the situation of a periodic Hill operator, we can show that the measure defined by the distribution function $N(\lambda)$ is absolutely continuous with respect to the Lebesque measure $d\lambda$.

Definition 1.4.10 The **density of states (DOS)** $\rho(\lambda)$ is defined as the Radon–Nikodym derivative of that measure with distribution function $N(\lambda)$:

$$\rho(\lambda) := \frac{dN(\lambda)}{d\lambda}. \tag{1.4.10}$$

One can combine the relations between the functions introduced in this section as follows:

Theorem 1.4.11

$$N(\lambda) = \frac{1}{\pi} k(\lambda) = 2\omega(\lambda) \tag{1.4.11}$$

and thus

$$\rho(\lambda) = N'(\lambda) = \frac{k'(\lambda)}{\pi}. \tag{1.4.12}$$

Suppose now λ belongs to the spectrum, then, differentiating (1.4.1) and taking into account vanishing $k(\lambda)$ below the spectrum, we get

$$\rho(\lambda) = \frac{|\Delta'(\lambda)|}{\pi\sqrt{4 - \Delta^2(\lambda)}} \theta(4 - \Delta^2(\lambda)), \tag{1.4.13}$$

where θ is the Heaviside function.

If a is a boundary point of an open gap, then this function has a singularity of type $(\lambda - a)^{-1/2}$ on the spectral side. If the gap at a is closed – that is, the bands touch – the singularity disappears, as shown in Fig. 1.9.

1.5 Spectral Asymptotics

An important and actively studied question is the asymptotic behavior of the spectral quantities when $\lambda \to \infty$. The following results for the periodic Hill operator were obtained in Shenk and Shubin (1986, 1985) (see also Kappeler et al., 2012, and

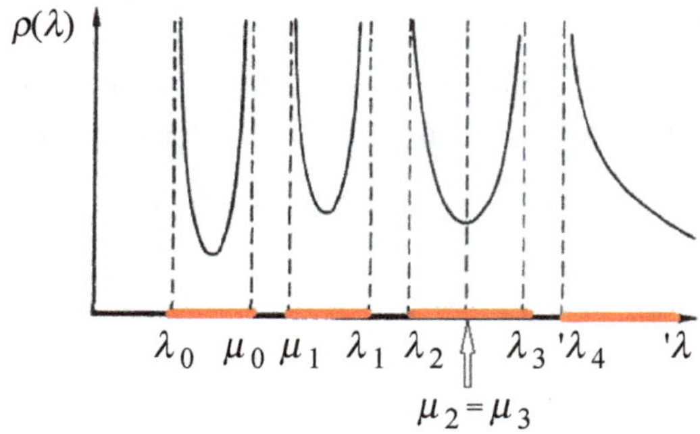

Figure 1.9 The behavior of the density of states $\rho(\lambda)$.

Section 17 of Rozenblyum et al., 1989, as well as Sobolev, 2003, 2006, for the case of pseudo-differential operators in dimension one).

Theorem 1.5.1 *Let the potential in the Hill equation be infinitely differentiable. Then the following asymptotic expansion holds when $\lambda \to \infty$: for any $N = 1, 2, \ldots$,*

$$e(\lambda; x, y) = \frac{\sin[\sqrt{\lambda}(x-y)]}{\pi(x-y)} + \sum_{k=1}^{N} \left\{ f_k(x, y) \int_\lambda^\infty \frac{\cos[\sqrt{\lambda}(x-y)]}{\lambda^{k+1/2}} d\lambda \right. \\ \left. + g_k(x, y) \int_\lambda^\infty \frac{\sin[\sqrt{\lambda}(x-y)]}{\lambda^k} d\lambda \right\} + O(\lambda^{-N}). \tag{1.5.1}$$

In short-hand notation,

$$e(\lambda; x, y) \sim \frac{\sin[\sqrt{\lambda}(x-y)]}{\pi(x-y)} + \sum_{k=1}^{\infty} \left\{ f_k(x, y) \int_\lambda^\infty \frac{\cos[\sqrt{\lambda}(x-y)]}{\lambda^{k+1/2}} d\lambda \right. \\ \left. + g_k(x, y) \int_\lambda^\infty \frac{\sin[\sqrt{\lambda}(x-y)]}{\lambda^k} d\lambda \right\}. \tag{1.5.2}$$

Moreover,

(1) *The expansion is uniform with respect to x, y so that $|x - y| < A$ for a fixed A.*
(2) *The functions f_k and g_k are independent of N and belong to $C^\infty(\mathbb{R} \times \mathbb{R})$.*
(3) *The asymptotic expansion can be differentiated arbitrarily many times with respect to x and y, if for some fixed m, λ belongs to the set*

$$M_m := \{\lambda \geq 1 \mid \text{dist}\,(\lambda, \mathbb{R} \setminus \sigma(H)) \geq \lambda^{-m}\}.$$

(4) *It can also be differentiated with respect to* λ. *When taking the derivative* $\frac{\partial^{p+r+s}}{\partial\lambda^p\partial x^r\partial y^s}$, *the term* $O(\lambda^{-N})$ *should be replaced with* $O(\lambda^{-N-p/2+(r+s)/2})$.
(5) *In particular, for* $x = y$, *one has uniform expansion on the whole* $\mathbb{R}$:

$$e(\lambda;x,x) \sim \pi^{-1}\sqrt{\lambda} + \sum_{k=0}^{\infty} h_k(x)\lambda^{-k-1/2},\ \lambda \to \infty. \tag{1.5.3}$$

Corollary 1.5.2

$$N(\lambda) \sim \pi^{-1}\sqrt{\lambda} + \sum_{k=0}^{\infty} d_k\lambda^{-k-1/2},\ \lambda \to \infty. \tag{1.5.4}$$

1.6 Localized Perturbations and Impurity Spectra

An important question is about what happens to the spectrum of the periodic Schrödinger operator when a sufficiently fast-decaying (e.g., compactly supported, or exponentially decaying, or just with finite first or second momentum) "perturbation" potential q is added:

$$Lu := -\frac{d^2}{dx^2} + V(x) + q(x), \tag{1.6.1}$$

where V is, as before, the periodic potential, and $q(x)$ is in some sense "small at infinity." Physically, this corresponds to inserting a localized impurity (e.g., "a wrong or missing atom") into a crystal. In particular, we will be interested in the following cases:[15]

(1) Integrability

$$q \in L_1(\mathbb{R}); \tag{1.6.2}$$

(2) Finiteness of the first moment:

$$\int_{\mathbb{R}} |q(x)|(1 + |x|)dx < \infty; \tag{1.6.3}$$

(3) Finiteness of the second moment:

$$\int_{\mathbb{R}} |q(x)|(1 + |x|^2)dx < \infty. \tag{1.6.4}$$

A well-known theorem by Weyl (1909) (see Theorem 5.35 of Kato, 1995) claims that under **relatively compact** additive **perturbations** (this means compact as multiplication operators in the operator norm of the operator $-d^2/dx^2$, which holds in all the cases listed earlier), the essential spectrum does not change, albeit some isolated eigenvalues might appear in the spectral gaps or even inside of

[15] In higher dimensions, when things get much harder, one often needs stronger decay conditions.

the continuous spectrum. The former condition was discussed in a renowned article by Von Neumann and Wigner (1929), where the case of zero background potential was considered.

We now collect the known results on the existence and the number of such impurity eigenvalues for the operator (1.6.1) (refer to comments and references discussed in Section 1.7):

Theorem 1.6.1

(1) *If* (1.6.2) *is satisfied, then the absolutely continuous spectrum of the perturbed operator is the same as that of the non-perturbed one. Eigenvalues cannot be embedded into the interior of any spectral band.*

(2) *If the first moment is finite, then the number of impurity eigenvalues in each spectral gap is finite, and for sufficiently far away gaps (i.e., for large* $|\lambda|$*), it does not exceed* 2.

(3) *If additionally the perturbation does not change sign, then there is at least one eigenvalue in each spectral gap, and sufficiently far in the spectrum (i.e., for large* $|\lambda|$*) – exactly one in each gap). More generally, this is true if the integral of* q *does not vanish.*

(4) *If the second moment is finite, then there is no impurity eigenvalue at any edge of the spectral bands. Additionally, there is at least one of them in any sufficiently far away spectral gap. If, additionally, the perturbation is bounded, then there is exactly one such eigenvalue in each sufficiently far away spectral gap.*

1.7 Comments and References

- **Section 1.1**: A much more sophisticated and hard to prove analog of this Euler's theorem 1.1.1 (L. Ehrenpreis called it the fundamental principle) also holds for constant coefficient linear PDEs (Ehrenpreis, 1970; Malgrange, 1955–1956, 1963; Palamodov, 1970). The corresponding analogs for periodic PDEs can be found in Kuchment (1993).
- **Section 1.6**: In the case of a 1D-perturbed periodic Schrödinger operator, counting the number of independent solutions shows the absence of these impurity eigenvalues embedded in the **interior** of a spectral band, which proves the first statement of Theorem 1.6.1. However, they are allowed to appear in the spectral gaps and even at their edges. The whole first claim was proven in Rofe-Beketov (1964a).

  The second and the fourth statements of the theorem were also proven in Rofe-Beketov (1964a), the third one in Želudev (1970, 1968, 1967).

Such results are also discussed in Rofe-Beketov (1973, 1977, 1979), see also Rofe-Beketov and Kholkin (2005, Ch. 6) and Brown et al. (2013, Section 5.3).

An analog of Theorem 1.6.1 holds (see Behrndt et al., 2023) for a more general Sturm–Liouville operator

$$\frac{1}{p(x)}\left(-\frac{d}{dx}q(x)\frac{d}{dx}+r(x)\right), \tag{1.7.1}$$

with periodic p, q, and r and non-vanishing p and q. Here one may perturb all three coefficients.

We will address the higher-dimensional case in Chapter 6. As the reader will understand the important issue of absence of embedded eigenvalues in higher dimensions becomes very tricky (e.g., such eigenvalues might appear) and is far from being completely resolved.

2

Multidimensional Periodicity: Lattices, Fundamental Domains, and Fourier Series

2.1 Introduction

In order to study periodic partial differential equations, it is first necessary to go through the fundamental basics of lattices. A more detailed introduction can be found in Skriganov (1985a).

For algebraically inclined people, we will use the term **lattice** in $\mathbb{R}^n$ defined as follows:

Definition 2.1.1 A **(Bravais) lattice** in $\mathbb{R}^n$ is a discrete, co-compact subgroup Γ of the additive group $\mathbb{R}^n$.

For the rest of us, a lattice consists of all integer linear combinations of n fixed linearly independent vectors (i.e., a basis) $\mathbf{a}_j \in \mathbb{R}^n$, $j = 1, \dots, n$. Thus, each lattice vector $\gamma \in \Gamma$ can be uniquely represented as follows:

$$\gamma = \sum_{j=1}^{n} \gamma^j \mathbf{a}_j, \qquad \gamma^j \in \mathbb{Z}. \tag{2.1.1}$$

We will call such a set of vectors $\{\mathbf{a}_j\}$ a **basis of the lattice**. We will also use the following notation for a vector $\gamma \in \Gamma$:

$$|\gamma| := \sum |\gamma^j|, \tag{2.1.2}$$

where γ^j are the coefficients in (2.1.1).

If the space $\mathbb{R}^n$ is equipped with an inner product and the corresponding Euclidean norm $\|\cdot\|$, and the chosen basis is orthonormal, we will also use the notation $\|\gamma\| := \sqrt{\sum |\gamma^j|^2}$ for the length of $\gamma \in \Gamma$. We will identify Γ with the corresponding group of shifts of $\mathbb{R}^n$.

Equivalently, any lattice Γ is the image of the integer lattice $\mathbb{Z}^n \subset \mathbb{R}^n$ under a nondegenerate linear transformation. In most (but not all) cases, the questions we consider will be invariant under such transformations. Thus, in most cases, there will be no significant harm if the reader assumes that $\Gamma = \mathbb{Z}^n$.

An important warning: To ensure clarity, the reader should note the following details. In physics, the term **lattice** is often used in a broader sense to denote any discrete subset of $\mathbb{R}^n$, which is invariant (periodic) with respect to the shifts by elements of a lattice Γ as defined earlier. In this context, the lattice Γ of periods is referred to as a **Bravais lattice**.

In this text, when referring to a "lattice," we always mean a Bravais lattice, that is, a group Γ.

In some cases, in an Euclidean (i.e., in presence of an inner product) setting, the following notions in Definition 2.1.2 will be handy.

Definition 2.1.2

- A lattice Γ is said to be **rational**, if there is a number r such that the dot products $\gamma_1 \cdot \gamma_2$ of any elements $\gamma_1, \gamma_2 \in \Gamma$ are rationally proportional to r. In other words, $(\gamma_1, \gamma_2)/r$ is rational for any $\gamma_1, \gamma_2 \in \Gamma$.
- A lattice Γ is said to have a **simple length spectrum**, if for $\gamma_1, \gamma_2 \in \Gamma$, the length equality $\|\gamma_1\| = \|\gamma_2\|$ implies $\gamma_1 = \pm\gamma_2$.

2.2 Fundamental Domains

Let Γ be a lattice in $\mathbb{R}^n$ generated by basis vectors $\mathbf{a}_j$, $j = 1, \ldots, n$. Consider the domain U (a parallelepiped) consisting of all linear combinations $\sum c^j \mathbf{a}_j$ such that all coefficients c^j belong to the interval $(-0.5, 0.5)$. This domain has the following properties:

(i) **it is disjoint from all its translations $U + \gamma$ by nonzero elements γ of Γ;**
(ii) **the shifts of the closure $\overline{U}$ of domain U by elements of Γ cover the whole space.**

Definition 2.2.1 Any domain with properties (i) and (ii) will be called a **fundamental domain** for the lattice Γ.

Remark 2.2.2 Besides the parallelepiped U described earlier, the lattice Γ has many other fundamental domains, so a fundamental domain is far from being uniquely defined.

For instance, here is another popular (in particular, in solid-state physics and crystallography) choice of fundamental domain, which assumes that some Euclidean metric is chosen on $\mathbb{R}^n$:

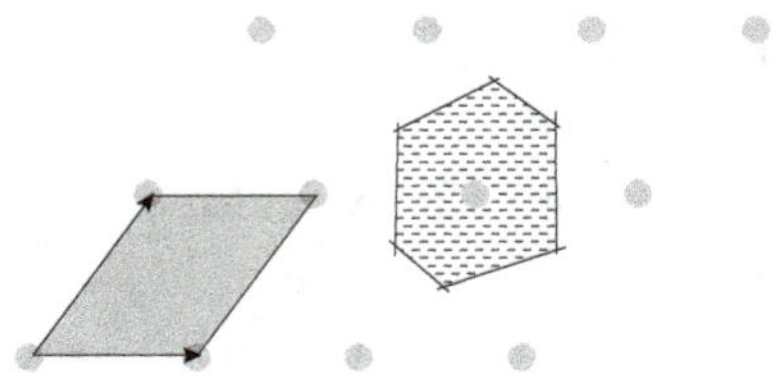

Figure 2.1 A lattice and two of its fundamental domains (shaded).

Definition 2.2.3 The **Voronoi cell** is the set W of all vectors $x \in \mathbb{R}^n$ that are closer to the origin than to any nonzero element of Γ.

The Voronoi cell is clearly a fundamental domain. An example of a Voronoi fundamental domain can be found on the right-hand side of Fig. 2.1.

Remark 2.2.4 Fundamental domains can be disconnected (one can construct simple examples even in 1D). In some cases, such "strange" domains might be useful. For instance, these include the so-called **higher Brillouin zones** in physics (Ashcroft and Mermin, 1976), which we will briefly describe now.[1]

Consider the Voronoi cell (temporarily call it "Brillouin zone") mentioned earlier: the set W of all vectors x in $\mathbb{R}^n$ that are closer to the origin than to any nonzero element of Γ. One can construct it geometrically by connecting the origin with all other points $\gamma \in \Gamma$ and then drawing perpendicular hyperplanes at the midpoints of the resulting line segments. Then, clearly, the Voronoi cell of the origin (the **first Brillouin zone** in physics parlance) consists of all points that are connected to the origin without crossing any of the hyperplanes (we disallow here going through intersections of the hyperplanes).

What if we consider the set of points that can be connected to the origin with exactly one intersection? This is the so-called **second Brillouin zone**. Increasing the number of allowed intersections, we define all **higher-order Brillouin zones** B_j, starting with $B_1 = W$.

Definition 2.2.5 The *m*th **Brillouin zone** consists of all points outside the union U of the Voronoi hyperplanes, which can be connected to the origin by crossing $(m - 1)$ (and no less) of these hyperplanes.

Figure 2.2 shows that the shapes of the iterated domains become increasingly complex, in particular, the domains can become disconnected.

The reader can consider as an exercise proving Proposition 2.2.6:

Proposition 2.2.6 *Each higher-order Brillouin zone is a fundamental domain.*

[1] We will return to Brillouin zones later on in this chapter, when considering the dual lattices.

Figure 2.2 Higher-order Brillouin zones (shown in different colors) for a hexagonal lattice in 2D.

Henceforth, we will consider, unless stated otherwise, only connected fundamental domains.

The following notion is customarily used in solid-state physics.

Definition 2.2.7 A fundamental domain for Γ (usually chosen as the Voronoi cell, although we will not insist on this) is known as the **Wigner–Seitz cell.**

Definition 2.2.8 If $\mathbf{W}$ is a fundamental domain, we denote its **volume** by v_Γ.

Since a fundamental domain is not uniquely defined, one might wonder whether v_Γ is also ambiguously defined? In fact – not.

Lemma 2.2.9 *The volume v_Γ is independent of the choice of a particular fundamental domain.*

The reader will see the explanation of why this is true in Section 2.2.1, when the torus $\mathbb{T} = \mathbb{R}^n/\Gamma$ will be introduced, and it will become clear that its volume is the same as that of *any* fundamental domain.

2.2.1 Torus

Let $\mathbf{W}$ be a fundamental domain for the lattice Γ. According to the definition, there are no points in $\mathbf{W}$ that differ by elements of Γ, while there must be such points on the boundary of $\mathbf{W}$. After identifying the points of the boundary of $\overline{\mathbf{W}}$ that differ by Γ-shifts, one obtains the **torus** $\mathbb{T} = \mathbb{R}^n/\Gamma = \overline{W}/\Gamma$. Here, we use the standard notation $\mathbb{R}^n/\Gamma$ for the set of orbits of action of the translation group Γ on $\mathbb{R}^n$, that is, points $x \in \mathbb{R}^n$ and $x + \gamma$ for $\gamma \in \Gamma$ are identified.

Remark 2.2.10

- It is usually convenient to consider Γ-periodic functions on $\mathbb{R}^n$ as functions on $\mathbb{T}$, which we will frequently do.
- Equipped with the operation induced by the addition of vectors in $\mathbb{R}^n$ and its natural topology, $\mathbb{T}$ is a **compact abelian (i.e., commutative) group**.
- If dx is the measure induced on $\mathbb{T}$ from $\mathbb{R}^n$, then $v_\Gamma^{-1} dx$ is the **Haar measure**[2] on $\mathbb{T}$.

Since the measure on $\mathbb{T}$ is induced from $\mathbb{R}^n$ by the quotient mapping $\mathbb{R}^n \mapsto \mathbb{T}$, one arrives at Lemma 2.2.11.

Lemma 2.2.11 *The quotient mapping $W \mapsto \mathbb{T}$ is onto and locally preserves the measure. In particular, the volumes of all fundamental domains are equal to the volume of the torus $\mathbb{T}$ and thus are all the same.*

2.3 Reciprocal Lattices

Working with periodic functions, one also needs the notion of the **dual (or reciprocal) lattice**. We will use the notations S^* for the dual to (i.e., the space of linear functionals on) the finite-dimensional space S and $\langle \phi, f \rangle$ for the value of $\phi \in S^*$ at $f \in S$.

Definition 2.3.1

- The vectors of the dual space $(\mathbb{R}^n)^*$ will be denoted as $\mathbf{k} = (k_1, \dots k_n)$ and called **quasi-momenta** (or **crystal momenta**), the name being derived from the solid-state physics.
- The **dual (reciprocal) lattice** $\Gamma^* \subset (\mathbb{R}^n)^*$ **to** Γ is defined as follows:

$$\Gamma^* = \{k \in (\mathbb{R}^n)^* \mid \langle k, \gamma \rangle \in 2\pi\mathbb{Z} \text{ for any } \gamma \in \Gamma\}. \tag{2.3.1}$$

- The original lattice Γ is sometimes called the **real lattice**.

If we assume that $\mathbb{R}^n$ is equipped with Euclidean structure, that is, with an inner product that will be denoted $\mathbf{a} \cdot \mathbf{b}$, then the dual space to $\mathbb{R}^n$ can be identified with $\mathbb{R}^n$ itself, and the definition then reduces to Definition 2.3.2

Definition 2.3.2 The **dual (or reciprocal) lattice** $\Gamma^* \subset \mathbb{R}^n$ (see Fig. 2.3) to the lattice $\Gamma \subset \mathbb{R}^n$ consists of all vectors $\mathbf{k} \in \mathbb{R}^n$ such that $\mathbf{k} \cdot \gamma \in 2\pi\mathbb{Z}$ for any $\gamma \in \Gamma$.

In particular, when $\Gamma = \mathbb{Z}^n$ and the duality is coming from the standard Euclidean scalar product, then $\Gamma^* = 2\pi\mathbb{Z}^n$.

[2] That is, the normalized invariant measure.

Figure 2.3 An example of a real lattice and its reciprocal.

Definition 2.3.3 We fix a fundamental domain $\mathcal{B}$ of (the reciprocal lattice) Γ^* in $(\mathbb{R}^n)^*$, which is known as the **Brillouin zone.**

One of the options (standard in physics) is to choose the Voronoi cell of $0 \in \Gamma^*$ (see Fig. 2.4). In this work, we'll be choosing a parallelepiped shown in Fig. 2.1 as the fundamental domain and also skip the consideration of higher Brillouin zones (see, e.g., Colin de Verdière, 1991; Skriganov, 1985a for discussions). We use notation v_{Γ^*} for the volume of the Brillouin zone $\mathcal{B}$.

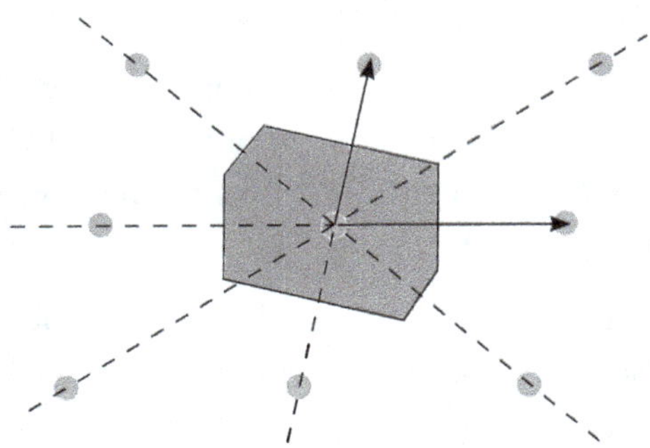

Figure 2.4 A Brillouin zone of a lattice.

Proposition 2.3.4 *The following equality holds:*

$$v_\Gamma v_{\Gamma^*} = (2\pi)^n. \tag{2.3.2}$$

Proof Since we know that the volumes v_Γ and v_{Γ^*} do not depend on the particular choice of fundamental domains, one can consider a parallelepiped one, in which case (2.3.2) can be easily proven. □

Remark 2.3.5 The particular choice of a lattice does not matter in many results (where there is invariance with respect to basis change) but does matter in other results. In simpler terms, we will assume that, unless noted otherwise,

$$\Gamma = \mathbb{Z}^n, \quad \Gamma^* = 2\pi\mathbb{Z}^n, \quad W = [0,1]^n, \quad \mathcal{B} = [-\pi,\pi]^n. \tag{2.3.3}$$

As in the case of the original lattice Γ, we can introduce the **dual torus** $\mathbb{T}^*$ that correspond to the reciprocal lattice:

$$\mathbb{T}^* := (\mathbb{R}^n)^*/\Gamma^* = \mathcal{B}/\Gamma^*. \tag{2.3.4}$$

Analogously to the torus $\mathbb{T}$, torus $\mathbb{T}^*$ is a compact abelian group equipped with normalized Haar measure $v_{\Gamma^*}^{-1}dk$, where dk is the Euclidean measure on $(\mathbb{R}^n)^*$.

Remark 2.3.6

- Under our assumption that $\Gamma = \mathbb{Z}^n$, the torus $\mathbb{T}^* = \mathbb{R}^n/2\pi\mathbb{Z}^n$ can be interpreted as the unit torus in $\mathbb{C}^n$:

$$\mathbb{T}^* = \{(z_1, \ldots, z_n) \in \mathbb{C}^n \mid |z_1| = \cdots = |z_n| = 1\}, \tag{2.3.5}$$

  where $z_j = e^{ik_j}$.
- This set is a part of the boundary of the unit polydisc,

$$D := \{(z_1, \ldots, z_n) \in \mathbb{C}^n \mid |z_j| < 1 \text{ for all } j = 1, \ldots, n\}.$$

  Having insufficient dimension, it is clearly not the whole boundary. However, it is known to be sufficient for having a maximum principle for analytic functions. It is called the **Shilov boundary** (also known as **Bergman–Shilov boundary**). See, for example, Rudin (1969).

This identification proves to be very useful, enabling powerful techniques from several complex variables, as the subsequent development of the theory will demonstrate.

2.3.1 Reduced Fundamental Domains

Frequently, additional symmetries are present in the space of quasi-momenta, besides the Γ^*-periodicity. For instance, as will be explained later, if the coefficients of the relevant operator (e.g., stationary Schrödinger) are real, then the central symmetry $k \to -k$ is present.[3] In this case, one does not have to work with the whole Brillouin zone but rather with any of its "halves" that does not contain pairs of symmetric points. For instance, if $\Gamma = \mathbb{Z}^n$, then instead of using the whole Brillouin zone $\mathbf{B} = [-\pi,\pi]^n$, one could use a reduced zone, that is $[0,\pi] \times [-\pi,\pi]^{n-1}$.

The operator of interest might also be symmetric with respect to an additional

[3] This symmetry reflects the time-reversal symmetry of dynamical equations and does not hold if the coefficients are not real, for example, in the presence of a magnetic potential, which breaks the time-reversal symmetry.

group of "**point symmetries**," that is, transformations that have fixed points.[4] For instance, it might commute with sign changes of coordinates or with a specific rotation. For example, in the case of commuting with sign changes, the analysis can be restricted to the reduced zone $[0, \pi]^n$.

Definition 2.3.7 A **reduced Brillouin zone** is a fundamental domain of the full Brillouin zone with respect to the action of all available point symmetries.

In other words, a reduced Brillouin zone is a subdomain U of a full zone $\mathbf{B}$, such that applying to U all point symmetries of the quasi-momentum space, one obtains the whole $\mathbf{B}$. For instance, if $\Gamma = \mathbb{Z}^2$ and the operator is symmetric with respect to the exchange of variables $x_1 \Leftrightarrow x_2$ and sign changes of each variable, one can use the triangle $x_1 \in [0, \pi], x_2 \in [0, x_1]$ as a reduced Brillouin zone.

Another comment that turns out to be of importance is that the presence of additional point symmetries implies the existence of **some special symmetry points** in the Brillouin zone, that is, points that are fixed by some of the symmetries. For instance, if $\Gamma = \mathbb{Z}^2$ and the time-reversal symmetry $k \to -k$ is present, then all points in the Brillouin zone $\mathbf{B} = [-\pi, \pi]^2$ that have coordinates equal to 0 or $\pm\pi$ are fixed by some of the symmetries. Indeed, if $k = -(k + k_0)$ for some $k_0 \in \Gamma^* = 2\pi\mathbb{Z}^2$, then $k \in \pi\mathbb{Z}^2$. The importance of these **high-symmetry points** in the Brillouin zone has been realized by physicists since the article Bouckaert et al. (1936) (see also Ashcroft and Mermin, 1976; Kittel, 1976). For some special lattice geometries that appear often in the solid-state theory, these symmetry points even have specific names assigned[5] (X, Γ, M, etc.) (see, e.g., Ashcroft and Mermin, 1976; Kittel, 1976). We will return to the discussion of the importance of these points in Section 4.6.2.

2.3.2 Periodic Functions and Fourier Series

Let $f(x)$ be a measurable function on $\mathbb{R}^n$ and $\Gamma \subset \mathbb{R}^n$ be a lattice.

Definition 2.3.8 We say that f is **Γ-periodic**, if $f(x + \gamma) = f(x)$ for any $\gamma \in \Gamma$ and a.e. with respect to x.

Periodicity can also be naturally defined for distributions.

Let us now recall some standard facts about multidimensional Fourier series (e.g., Stein and Weiss, 1971). Let $f(x) \in L^2_{\text{loc}}(\mathbb{R}^n)$ be Γ-periodic. The space of such functions can be naturally identified with $L^2(\mathbb{T})$, where, as before, $\mathbb{T}$ is the torus $\mathbb{R}^n/\Gamma$. The Lebesgue measure on $\mathbb{R}^n$ produces a (non-normalized) invariant

[4] The full symmetry group, called **space, or crystallographic group** (Ashcroft and Mermin, 1976; Kittel, 1976; Knox and Gold, 1964; Slater, 1972), is generated by the point and shift symmetries and is usually noncommutative.

[5] We will not use these notations.

measure on $\mathbb{T}$, with respect to which the total measure of the torus is equal to v_Γ. For any $k \in \Gamma^*$, the function $\exp(ik \cdot x)$ belongs to $L^2(\mathbb{T})$. These functions are pairwise orthogonal, form a complete family in $L^2(\mathbb{T})$, and have norms equal to $v_\Gamma^{1/2}$. Thus, the functions $v_\Gamma^{-1/2} e^{ik\cdot x}$ form an orthonormal basis in this space. Now one can define in the standard way the **Fourier series** of any function $f \in L^2(\mathbb{T})$:

$$f \rightarrowtail v_\Gamma^{-1/2} \sum_{k\in\Gamma^*} f_k e^{ik\cdot x}, \tag{2.3.6}$$

where the **Fourier coefficients** f_k are computed as follows:

$$f_k = v_\Gamma^{-1/2} \int_{\mathbb{T}} f(x) e^{-ik\cdot x} dx. \tag{2.3.7}$$

Due to the orthonormality and completeness of the system, the Plancherel theorem holds. That is, the mapping $f \to \{f_k\}$ is an isometry between $L^2(\mathbb{T})$ and the space $l_2(\Gamma^*)$ of square-summable functions on Γ^*.

It is easy to check, going through the standard proofs, that the same holds for functions with values in a Hilbert space.

Theorem 2.3.9 *Let H be a complex Hilbert space. Then the relations* (2.3.8) and (2.3.9) *(Hilbert space analogs of* (2.3.6) and (2.3.7)*) define an isometry between the space $L^2(\mathbb{T}, H)$ of measurable square-summable H-valued functions on the torus $\mathbb{T}$, and the space $l_2(\Gamma^*, H)$ of square-summable H-valued functions on the reciprocal lattice Γ^*.*

Remark 2.3.10 From the point of view of the representation theory (e.g., Barut and Râczka, 1986; Simon, 1996; Želobenko, 1973, 2006), the discrete abelian group Γ^* is the dual group of the compact abelian group $\mathbb{T}$. The functions $v_\Gamma^{-1} e^{ik\cdot x}$ when $k \in \Gamma^*$ (**unitary characters**) provide all irreducible unitary representations of $\mathbb{T}$, and the mapping $f \to \{f_k, k \in \Gamma^*\}$ is the corresponding (isometric) Fourier transform.

The relation between $\mathbb{T}^*$ and Γ is the same as that between $\mathbb{T}$ and Γ^*. Hence, a similar construction applies: the mappings between $L^2(\mathbb{T}^*, H)$ and $l_2(\Gamma, H)$, that is,

$$g(k) \rightarrowtail v_{\Gamma^*}^{-1/2} \sum_{\gamma\in\Gamma} g_\gamma e^{ik\cdot\gamma} \tag{2.3.8}$$

and

$$g_\gamma = v_{\Gamma^*}^{-1/2} \int_{\mathbb{T}^*} g(k) e^{-ik\cdot\gamma} dk, \tag{2.3.9}$$

are mutually inverse and isometric.

In what follows, we will often need to look at this from a different perspective.

Specifically, one can consider the mapping from Fourier coefficients to the corresponding periodic functions as the Fourier transform on a discrete abelian group. Let us have a square-summable H-valued function f_γ on Γ, that is, a sequence $\{f_\gamma\}, \gamma \in \Gamma$ that belongs to $l_2(\Gamma, H)$. Then its *Fourier transform* is the function

$$\widetilde{f}(k) = v_{\Gamma^*}^{-1/2} \sum_{\gamma\in\Gamma} f_\gamma e^{ik\cdot\gamma} \in L^2(\mathbb{T}^*, H), \tag{2.3.10}$$

which has the norm equal to the one of the original. An analogous correspondence can be made between $l_2(\Gamma^*, H)$ and $L^2(\mathbb{T}, H)$, although we will mainly not need it.

Theorem 2.3.9 can be considered as the Plancherel theorem for the group Fourier transforms indicated earlier. Sometimes analogs of the Paley–Wiener theorem are also needed. We mean by this statements that describe the spaces of periodic functions whose Fourier coefficients decay in a certain prescribed way.

We will start by considering the condition for a superpolynomial decay.

Definition 2.3.11 Let H be a complex Hilbert space as defined earlier. The (Schwartz type) space $\mathcal{S}(\Gamma, H)$ consists of all H-valued functions $\{f_\gamma\}$, $\gamma \in \Gamma$ on Γ such that, for any $N > 0$, the expression

$$\rho_N(\{f_\gamma\}) = \sup_{\gamma\in\Gamma}\{ \|f_\gamma\|_H |\gamma|^N\} \tag{2.3.11}$$

is finite.

The collection of semi-norms ρ_N defines a Fréchet space topology on $\mathcal{S}(\Gamma, H)$.

The following simple statement is the vector-valued analog of the classical theorem of Fourier analysis (e.g., Stein and Weiss, 1971): Super-polynomial decay of Fourier coefficients is equivalent to infinite smoothness of the periodic function. Its proof is parallel to that of the scalar case (see, e.g., Theorems 2.2.2–2.2.5 and Lemma 3.1.2 of Kuchment, 1993).

Theorem 2.3.12 *The Fourier transform $\{f_\gamma\} \to \widetilde{f}(k)$ is a topological isomorphism of the space $\mathcal{S}(\Gamma, H)$ onto the space $C^\infty(\mathbb{T}^*, H)$ of infinitely differentiable H-valued functions on the torus $\mathbb{T}^*$, equipped with the natural Fréchet topology.*

If we require exponential decay of Fourier coefficients, this results in analyticity of the corresponding periodic function. In order to formulate this statement precisely, we need to introduce some notation first.

Definition 2.3.13 Let for any $a > 0$, the space $A_a(\Gamma, H)$ consist of all H-valued functions f_γ on Γ such that, for any $b \in (0, a)$, the following expression is finite:

$$\psi_b(f) = \sup_{\gamma\in\Gamma} \|f_\gamma\|_H e^{b|\gamma|}. \tag{2.3.12}$$

This space is equipped with the Fréchet topology defined by the semi-norms ψ_b.

Definition 2.3.14 Let $\mathcal{D} \subset \mathbb{C}^n$ be an open domain. The space $A(\mathcal{D}, H)$ consists of all H-valued analytic functions in $\mathcal{D}$. The space is equipped with the Fréchet topology of uniform convergence on compact subsets of $\mathcal{D}$.

For any $0 < a \leq \infty$, we define the domain $\mathcal{D}_a \subset \mathbb{C}^n$ as follows:

$$\mathcal{D}_a = \{k \in \mathbb{C}^n \mid |\mathrm{Im}\, k \cdot \mathbf{a}_j| < a, \quad j = 1, \ldots, n\}, \tag{2.3.13}$$

where $\{\mathbf{a}_j\}$ is the basis of the lattice Γ mentioned earlier. This is just a tubular neighborhood (in l_∞ distance) of the real subspace.

We are ready now to introduce an analog of the standard Paley–Wiener theorem (see, e.g., Theorem 2.2.2 of Kuchment, 1993).

Theorem 2.3.15 *For any $a \in (0, \infty]$, the Fourier transform $\{f_\gamma\} \to \widetilde{f}(k)$ is a topological isomorphism of the space $A_a(\Gamma, H)$ in the subspace of Γ^*-periodic functions in $A(\mathcal{D}_a, H)$.*

We would like to mention now that something awkward was going on with the transform (2.3.10) and the further discussion. Indeed, the function defined in (2.3.10) is Γ^*-periodic, and hence, it is claimed to be a function on the torus $\mathbb{T}^*$. However, the variable k lives in $\mathbb{R}^n$. In other words, the mapping from $k \in \mathbb{R}^n$ to irreducible representations $e^{ik\cdot\gamma}$ of Γ is not one-to-one. It is easy to introduce another variable z that will reside on $\mathbb{T}^*$ and in terms of which the irreducible representations of Γ enter in a unique way. Let, as before, $\mathbf{a}_j$ be a basis of Γ and $\mathbf{b}_j$ be its biorthogonal basis, which is hence also a basis of Γ^*. Let us also employ the notation (2.1.1) $\gamma = \sum \gamma^j \mathbf{a}_j$. Consider the vector

$$z = (z_1, \ldots, z_n) \in \mathbb{C}^n, z_j = e^{ik\cdot\mathbf{a}_j}, \quad j = 1, \ldots, n.$$

Obviously, $|z_j| = 1$ for all j, so z belongs to the unit torus in $\mathbb{C}^n$. We will use the same notation $\mathbb{T}^*$ for this torus. The mapping $k \mapsto z$ is Γ^*-periodic. Thus, the characters $e^{ik\cdot\gamma}$ can be rewritten in a unique way as $z^\gamma := z_1^{\gamma^1} \cdots z^{\gamma^n}$, where the correspondence between the (unitary) characters and points $z \in \mathbb{T}^*$ is now one-to-one.

Using the same formula, this correspondence can be extended to non-unitary characters (i.e., to complex values of k) as well. Then the image of the whole space $\mathbb{C}^n$ of values of k will be the domain $\Omega := (\mathbb{C} \setminus \{0\})^n \subset \mathbb{C}^n$. The following domains will now replace the domains $\mathcal{D}_a$:

$$\Omega_a := \{z = (z_1, \ldots, z_n) \in \Omega \mid e^{-a} < |z_j| < e^a, \quad j = 1, \ldots, n\}.$$

We will abuse notations once more denoting the function $\widetilde{f}(k)$, written in z-variables, by $\widetilde{f}(z)$. The Plancherel and Paley–Wiener Theorems 2.3.9, 2.3.12, and 2.3.15 can be now equivalently restated in these new notations.

Theorem 2.3.16 *Let H be a complex Hilbert space. Consider the Fourier transform $\{f_\gamma, \gamma \in \Gamma\} \mapsto \widetilde{f}(z)$, then*

(1) *It maps isometrically $l_2(\Gamma, H)$ onto $L^2(\mathbb{T}^*, H)$.*
(2) *It maps isomorphically the Fréchet space $\mathcal{S}(\Gamma, H)$ onto $C^\infty(\mathbb{T}^*, H)$.*
(3) *For any $a \in (0, \infty]$, it maps isomorphically the Fréchet space $A_a(\Gamma, H)$ onto the space $A(\Omega_a, H)$ of holomorphic H-valued functions in Ω_a (this space being equipped with topology of uniform convergence on compacta).*

3

Floquet Transform and Direct Integral Decomposition

3.1 Introduction

Well, it is finally time to think about transplanting the Floquet theory to PDEs.

Our main 1D tools were the spectral analysis of the monodromy operator and the Lyapunov reduction theorem. Both of them rely on the propagator along the solutions of the Cauchy problem. This raises some hopes that an analog of the 1D Floquet theory **might** work for time-periodic PDEs of evolution type:

$$u_t = A(t)u \tag{3.1.1}$$

(parabolic, hyperbolic, non-stationary Schrödinger), where such propagators are nicely defined. Indeed, **sometimes** this does happen, although, surprisingly, some basic issues remain unresolved. For example, even the existence of (analogs of) Floquet solutions has not been proven for the heat equation

$$u_t = \Delta u + b(x,t)u, \tag{3.1.2}$$

where x belongs to a smooth-bounded domain, the smooth function $b(x,t)$ is periodic in t, and Dirichlet or Neumann conditions are imposed on the boundary. Some spectacular (but very restrictive) positive results for special cases have been obtained (Chow et al., 1994, 1995); see also surveys and further references in Kuchment (1993, 2016). The author plans to address this and other issues in Kuchment (2027).

However, for non-evolution (e.g., elliptic) periodic PDEs, which are our main objects of interest, using such propagators is all but impossible.[1]

Besides apparent impossibility of working with monodromy operators, analogs of some results of the Floquet theory for ODEs just do not hold for elliptic equations. For instance, there is no hope for an analog of the Lyapunov theorem 1.2.8. Indeed, such an analog would imply that the geometry of the set of Floquet multipliers (or

[1] Albeit such an approach was successfully adopted to some problems of the theory of periodic waveguides, where the governing equation is elliptic and periodic in one variable (see Derguzov, 1972a, 1975, as well as quite a few other articles by this author), one still realizes how limited is its power.

quasi-momenta) would be essentially the same for periodic and constant coefficient problems, which is not the case, as it will be discussed in Chapter 4.

One thus has to adopt Floquet theory technique for the higher spatial dimensions, which is the topic of this chapter.

In Sections 3.2–3.4, we introduce and study properties of the long-promised Floquet transform. A modification that is sometimes useful is provided in Section 3.5. Section 3.6 discusses the effect the transform has on periodic operators. In Section 3.7, functions of Bloch and Floquet types, which are analogs of exponents and exponential polynomials, are introduced and studied. Section 3.8 concentrates on dual interpretations of Floquet–Bloch functions as solutions of periodic elliptic equations and on counting dimensions of such solutions. For readers' convenience, the chapter ends (as several other chapters of the book do) with comments and references in Section 3.9, summarizing the chapter's main notions and results, as well as other comments and references.

3.2 Floquet Transform

We aim at equations that involve linear partial differential operators in $\mathbb{R}^n$ with coefficients periodic with respect to a lattice Γ. In other words, action of the operator on functions on $\mathbb{R}^n$ commutes with shifts of functions by elements of Γ. How can one use this information to simplify studying the equation? The general ideology of harmonic analysis (Mackey, 1978) gives the answer: One should use the Fourier transform on Γ. We already know how to do so for functions defined on Γ (see Chapter 2). However, functions $f(x)$ that we will be interested in are defined on the whole space $\mathbb{R}^n$ rather than on Γ. The situation is resolved if one defines the **shifted functions** $f_\gamma(x) = f(x - \gamma)$ and considers f_γ as a (function-valued) function of $\gamma \in \Gamma$.

As recalled in Chapter 2, Fourier series transform functions of $\gamma \in \Gamma$ (even with values in a Hilbert space) into functions of $z \in \mathbb{T}^*$:

$$\{f_\gamma\}_{\gamma\in\Gamma} \mapsto \widehat{f}(z) := \sum_{\gamma\in\Gamma} f_\gamma z^\gamma = \sum_{\gamma\in\Gamma} f_\gamma e^{ik\cdot\gamma},$$

with standard L^2 isometry and Paley–Wiener-type theorems being preserved. Here $k \in \mathbb{R}^n$ and z belongs to the unit torus $\mathbb{T}^*$.

Let now $f(x)$ be a (say, compactly supported for starters) function in $L^2(\mathbb{R}^n)$. We cover $\mathbb{R}^n$ by the shifted copies of the fundamental domain W:

$$\mathbb{R}^n = \bigcup_{\gamma\in\Gamma} (W - \gamma), \tag{3.2.1}$$

cut f into the corresponding pieces $f|_{(W-\gamma)}$ and then shift them all back to the original W:

$$f_\gamma(x) := f|_{(W-\gamma)}(x-\gamma). \tag{3.2.2}$$

If we restrict here x to being in the Wigner–Seitz cell W, we get a function on Γ with values in the Hilbert space $L^2(W)$:

$$\gamma \in \Gamma \mapsto f_\gamma \in L^2(W). \tag{3.2.3}$$

Now summing the Fourier series, we arrive at our main tool, Definition 3.2.1.

Definition 3.2.1 The **Floquet transform** [2] $\mathcal{U}_\Gamma$ acts as follows:

$$f(x) \mapsto (\mathcal{U}_\Gamma f)(x,k) := \widehat{f}(x,k) := \sum_{\gamma\in\Gamma} f_\gamma(x)e^{ik\cdot\gamma} = \sum_{\gamma\in\Gamma} f_\gamma(x)z^\gamma. \tag{3.2.4}$$

That is, this is just the sum of the Fourier series, whose coefficients are the pieces of f over the shifted copies of the fundamental domain.

Here, as before, $z = (z_1,\dots,z_n) := (e^{ik_1},\dots,e^{ik_n})$ is the Floquet multiplier corresponding to the **quasi-momentum (crystal momentum)** $k = (k_1,\dots,k_n)$. We will also abuse notations, writing

$$\widehat{f}(x,z) = \sum_{\gamma\in\Gamma} f_\gamma(x)z^\gamma.$$

It will be convenient to extend $\mathcal{U}_\Gamma f(x,k)$ from $x \in W$ to the whole $\mathbb{R}^n$ just by removing the restriction to $(W-\gamma)$ in (3.2.2), which results in

$$f(x) \mapsto \widehat{f}(x,k) := \sum_{\gamma\in\Gamma} f(x-\gamma)e^{ik\cdot\gamma} = \sum_{\gamma\in\Gamma} f(x-\gamma)z^\gamma. \tag{3.2.5}$$

If this is done, then one observes the following useful properties:

$$\widehat{f}(x+\gamma,k) = \widehat{f}(x,k)e^{ik\cdot\gamma}, \quad \gamma\in\Gamma, \tag{3.2.6}$$

$$\widehat{f}(x+\gamma,z) = \widehat{f}(x,k)z^\gamma, \quad \gamma\in\Gamma, \tag{3.2.7}$$

$$\widehat{f}(x,k+k') = \widehat{f}(x,k), \quad k'\in\Gamma^*. \tag{3.2.8}$$

The first two equalities, (3.2.6) and (3.2.7), say that, being considered on the whole space $\mathbb{R}^n$, function $\widehat{f}(x,k) = \mathcal{U}_\Gamma f(x,k)$ is **Γ-automorphic** with respect to x with

[2] Neither the name "Floquet transform" is commonly accepted, nor did Floquet introduce this transform. Various other names have been suggested: Bloch transform, Gel'fand transform (due to Gel'fand, 1950), Zak transform, and probably some others. Here, we see an instance of the so-called **Arnold's principle**: *If a notion bears a personal name, then this name is not the name of the discoverer* (Arnol'd, 1998). (See also **M. Berry's Corollary**: *the Arnold Principle is applicable to itself.*) Indeed, **Stigler's law of eponymy** (Stigler, 1980) had been known before and completely appropriately had not been proposed by Stigler. However, it is handy to use a name, rather than point a finger at a formula. Thus, the author chose Floquet, since this transform is very natural for Floquet theory.

the character $e^{ik\cdot\gamma}$ (this is also called **cyclic property, quasi-periodicity, or Floquet property**).[3] The last one, (3.2.8), claims its Γ^***-periodicity** with respect to k.

3.2.1 Inversion

To simplify many formulas, let us adopt the following definition.

Definition 3.2.2 We denote by $d'x$ (correspondingly $d'z$ or $d'k$) the **normalized Haar measure** $(v_\Gamma)^{-1}dx$ (correspondingly $(v_{\Gamma^*})^{-1}dz$ or $(v_{\Gamma^*})^{-1}dk$) on $\mathbb{T}$ (correspondingly $\mathbb{T}^*$).

It is easy to get an inversion formula for the Floquet transform. Indeed, it is enough to find the Fourier coefficients of $\mathcal{U}_\Gamma f(x,k)$ with respect to k and place them where they belong, that is, on the shifted copies of the fundamental domain. This leads to the following inversion formula:

$$f(x) = \int_{\mathbb{T}^*} \widehat{f}(x,k)e^{-ik\cdot\gamma}d'k, \qquad x \in W - \gamma, \text{ for any } \gamma \in \Gamma. \tag{3.2.9}$$

On the other hand, if we consider the Fourier coefficients f_γ as the shifted copies of f *in the whole space* $\mathbb{R}^n$ (rather than restricted to the Wigner–Seitz cell), then just the "constant" term of the Fourier series recovers the whole function:

$$f(x) = \int_{\mathbb{T}^*} \widehat{f}(x,k)d'k, \quad x \in \mathbb{R}^n. \tag{3.2.10}$$

Remark 3.2.3 The inversion formula (3.2.10) looks simpler than (3.2.9). It, however, has the disadvantage of using values of $\hat{f}(k,x)$ **for all** $x \in \mathbb{R}^n$, while in Floquet transform approach, we would like to use only $x \in \mathbf{W}$.

3.3 Quasi-momenta versus Floquet Multipliers

The periodicity property (3.2.8) of Floquet transform suggests that it should be possible to factor out the periodicity with respect to k by writing the resulting function $\hat{f}(k,x)$ as a function of the Floquet multiplier variable z instead of k, since z depends on k in a Γ^*-periodic manner. This can be easily done. Indeed, for real k, we can consider z as the element of the torus $\mathbb{T}^* = \mathbb{R}^n/\Gamma^*$ that corresponds to k under the "exponential" mapping $\mathbb{R}^n \overset{\Gamma^*}{\mapsto} \mathbb{T}^*$. This construction can be generalized for complex values of k. Namely, since Γ^* is a subgroup of the complexification $\mathbb{C}^n$ of $\mathbb{R}^n$, one can define an n-dimensional complex analytic manifold Ω as $\Omega = \mathbb{C}^n/\Gamma^*$ with the corresponding exponential mapping $\mathbb{C}^n \overset{\Gamma^*}{\mapsto} \Omega$.

[3] To put this simply, $f(x,k)$ is an eigenfunction of the shift operators with the eigenvalues $e^{ik\cdot x}$.

In the simplest case, when $\Gamma = \mathbb{Z}^n$ and $\Gamma^* = 2\pi\mathbb{Z}^n$, one can identify the torus $\mathbb{T}^*$ with the already familiar unit torus in $\mathbb{C}^n$

$$\mathbb{T}^* = \{z = (z_1, \dots, z_n) \in \mathbb{C}^n \mid |z_j| = 1, \quad j = 1, \dots, n\} \tag{3.3.1}$$

and the manifold Ω with

$$\Omega = (\mathbb{C} \setminus \{0\})^n = \{z = (z_1, \dots, z_n) \in \mathbb{C}^n \mid z_j \neq 0,\ j = 1, \dots, n\} \subset \mathbb{C}^n. \tag{3.3.2}$$

In this case, the relation between k and z is truly exponential:

$$z = e^{ik} := (e^{ik_1}, \dots, e^{ik_n}).$$

The reader can see that these definitions exactly match the ones for the ODE case (see Chapter 1).

Remark 3.3.1 Let

$$D := \{z = (z_1, \dots, z_n\} \in \mathbb{C}^n \ : \ \text{for all } j, |z_j| \leq 1\}$$

be the **unit polydisc**, an important domain in the several complex variables theory (Rudin, 1969). The torus $\mathbb{T}^*$, while not the whole boundary of D (having dimension too small) is, as we have mentioned earlier, the so-called **Shilov (or Bergman–Shilov) Boundary** (see, e.g., Agranovskiĭ, 1993; Grigoryan and Tonev, 2006; Krantz, 1999) of D.

For an arbitrary lattice Γ, one can analogously define the multipliers and the correspondence between them and quasi-momenta. Namely, let, as before, $\{\mathbf{a}_j\}$ be a basis of Γ.

Definition 3.3.2 The **Floquet multiplier** z corresponding to a *quasi-momentum* $k \in \mathbb{C}^n$ is defined as follows:

$$z = \left(e^{ik\cdot\mathbf{a}_1}, \dots, e^{ik\cdot\mathbf{a}_n}\right) \in \Omega := (\mathbb{C} \setminus \{0\})^n. \tag{3.3.3}$$

Under this definition, the representations (3.3.1) and (3.3.2) for the torus $\mathbb{T}^*$ and domain Ω still hold for a general lattice Γ.

Abusing notations, we will write the function $\hat{f}(k,x)$ as $\hat{f}(z,x)$, whenever we would like to consider it as a function of the Floquet multiplier z rather than of the quasi-momentum k.

One notices that Floquet multipliers might be more natural variables for Floquet-transformed functions rather than quasi-momenta. Indeed, unlike the quasi-momenta, they are uniquely defined. This is, however, only a relative advantage. Namely, as we will see later, using quasi-momenta one arrives to an operator polynomial of k acting on a fixed function space (of functions on $\mathbb{T}$), while in terms of

Floquet multipliers, one has a fixed operator restricted to an analytically "rotating" with z subspace. Thus, in some problems, it is to our advantage to stick with quasi-momenta, while in others, it will be more beneficial to go to the multiplier z variables.

Remark 3.3.3 We will need to use the quasi-momenta domains $\mathcal{D}_a$ introduced in (2.3.13). Their images in the space $\Omega = (\mathbb{C} \setminus \{0\})^n$ of Floquet multipliers will be denoted by Ω_a:

$$\Omega_a = \{z = (z_1, \ldots, z_n) \in \Omega \,|\, e^{-a} < |z_j| < e^a, \ j = 1, \ldots, n\}. \tag{3.3.4}$$

One can notice that domains Ω_a form a basis of neighborhoods of the torus $\mathbb{T}^* \subset \Omega$. In particular, $\Omega_\infty = \Omega$ (see 3.3.2).

3.4 Range Description and Plancherel and Paley–Wiener Theorems for Floquet Transform

3.4.1 Non-smooth Case

Let us try first to understand how to apply the Floquet transform to functions without compact support.

Assume that $f \in L^2(\mathbb{R}^n)$. Then one has

$$\|f\|^2_{L^2(\mathbb{R}^n)} = \sum_{\gamma \in \Gamma} \|f\|^2_{L^2(\mathbf{W}+\gamma)}. \tag{3.4.1}$$

This observation, together with the fact that we are interested in the values of $\hat{f}(k,x)$ for $x \in \mathbf{W}$ only, helps us to define the transform $\mathcal{U}_\Gamma$ for functions in $L^2(\mathbb{R}^n)$ and to prove the corresponding Plancherel-type theorem.

Theorem 3.4.1

(1) *If $f \in L^2(\mathbb{R}^n)$ and $K \subset \mathbb{R}^n$ is compact, then the series* (3.2.4) *converges in the space $L^2(\mathbb{T}^*, L^2(K))$.*

(2) *The following equality (Plancherel's theorem) holds:*

$$\|f\|^2_{L^2(\mathbb{R}^n)} = \int\limits_{\mathbf{B}} \|\hat{f}(k,\cdot)\|^2_{L^2(\mathbf{W})} d'k = \int\limits_{\mathbb{T}^*} \|\hat{f}(z,\cdot)\|^2_{L^2(\mathbf{W})} d'z, \tag{3.4.2}$$

where $\mathbf{W}$ *is the Wigner–Seitz cell of* Γ *and the prime in* d' *indicates using the normalized measure.*

Proof Let f and K be as in the theorem's statement. Consider the sequence of functions $f_\gamma \in L^2(K)$ defined as follows:

$$f_\gamma(x) = f(x+\gamma), x \in K, \gamma \in \Gamma.$$

Since the function f belongs to $L^2(\mathbb{R}^n)$, the sequence of numbers $\|f_\gamma\|_{L^2(\mathbf{W})}$ is square summable, that is, belongs to $L^2(\Gamma, L^2(\mathbf{W}))$. Now Theorem 2.3.9 applied to (2.3.10) implies the convergence statement, as well as the equality

$$\sum_{\gamma\in\Gamma} \|f_\gamma\|^2_{L^2(K)} = v_{\Gamma^*}^{-1} \int_{\mathbf{B}} \|\hat{f}(k,\cdot)\|^2_{L^2(K)} dk = \int_{\mathbf{B}} \|\hat{f}(k,\cdot)\|^2_{L^2(K)} d'k$$
$$= \int_{\mathbb{T}^*} \|\hat{f}(z,\cdot)\|^2_{L^2(K)} d'z. \tag{3.4.3}$$

Considering the particular case $K = \mathbf{W}$ and applying (3.2.2), one obtains (3.4.2). □

The main statement of this theorem is usually formulated equivalently as the expansion of the space $L^2(\mathbb{R}^n)$ into the following **direct integral**:

$$L^2(\mathbb{R}^n) = \int_{\mathbf{B}}^{\oplus} L^2(\mathbf{W})\, d'k = \int_{\mathbb{T}^*}^{\oplus} L^2(\mathbf{W})\, d'z. \tag{3.4.4}$$

The crude idea of a **direct integral** is the space of "measurable" functions of a parameter (k or z in our case) with values in Hilbert spaces (**fibers**) that also "nicely" depend upon the parameter. We will also use this notion sometimes, albeit we will not need the general theory of direct integrals of Hilbert spaces (Dixmier, 1957; Nielsen, 1980). The reason is that as we will see, the direct integral expansions that we need belong to a particularly nice class. In this class, the fibers "rotate analytically" (in technical terms, they form an **analytic Banach vector bundle**, see, for example, Zaidenberg et al., 1975, and Appendix D), and the operators we will be interested in also behave analytically with respect to the parameter.[4]

Abusing notations, we will also write

$$\int_{\mathbb{T}^*}^{\oplus} L^2(\mathbf{W})\, d'k,$$

although formally speaking quasi-momenta k do not live on the torus $\mathbb{T}^*$, but rather the Floquet multipliers z do. We might use this abuse sometimes (well, in fact quite often), when this cannot lead to errors. As a vigilance exercise for the reader, we will do this right away, in the next statement.

Having the previous theorem available and applying Theorem 2.3.16 to $\mathcal{H} = L^2(W)$ leads to the following **range theorems for the Floquet transform**:

Theorem 3.4.2

- $\mathcal{U}_\Gamma$ *is an isometry from* $L^2(\mathbb{R}^n)$ *onto* $L^2(\mathbb{T}^*, L^2(W))$.

[4] This situation would definitely change if we had to deal with a noncommutative group of periods.

- $\mathcal{U}_\Gamma f(\cdot, k)$ *is infinitely differentiable on* $\mathbb{T}^*$ *as a function with values in* $L^2(W)$ *if and only if the norms* $\|f\|_{L^2(W+\gamma)}$ *decay when* $|\gamma| \mapsto \infty$ *faster than any power of* $|\gamma|$.
- $\mathcal{U}_\Gamma f(\cdot, k)$ *is analytic in a complex neighborhood of* $\mathbb{T}^*$ *as a function with values in* $L^2(W)$, *if and only if the norms* $\|f\|_{L^2(W+\gamma)}$ *decay exponentially fast when* $|\gamma| \mapsto \infty$.

We will obtain now a Paley–Wiener-type theorem for Floquet transform (still assuming no smoothness of the original function). Let $a \in (0, \infty]$. Consider the following Fréchet space consisting of exponentially decaying (in L^2-sense) functions:

Definition 3.4.3 The space $L^{2,a}(\mathbb{R}^n)$ consist of all functions $f \in L_{2,\mathrm{loc}}(\mathbb{R}^n)$ such that for any $b \in (0, a)$, the following expression is finite:

$$\psi_b(f) = \sup_{\gamma\in\Gamma} \|f\|_{L^2(\mathbf{W}+\gamma)} e^{b|\gamma|}. \tag{3.4.5}$$

This space is equipped with the Fréchet topology defined by the semi-norms ψ_b.

Remark 3.4.4 It is useful to notice that the expression

$$\sup_{\gamma\in\Gamma} \|f\|_{L^2(K+\gamma)} e^{b|\gamma|} \tag{3.4.6}$$

is finite for any function $f \in L^{2,a}(\mathbb{R}^n)$, any compact K in $\mathbb{R}^n$, and $b \in (0, a)$.

Now Theorem 2.3.15 shows that the following Paley–Wiener-type theorem holds:

Theorem 3.4.5 *For any* $a \in (0, \infty]$, *the Floquet transform*

$$\mathcal{U}_\Gamma : f \mapsto \hat{f}(z, x)$$

is a topological isomorphism of the space $L^{2,a}(\mathbb{R}^n)$ *onto* $A(\Omega_a, L^2(\mathbf{W}))$.

In this theorem, we assumed that the Floquet transform is presented in terms of Floquet multipliers z, rather than quasi-momenta k (the trace of this assumption can be seen in the appearance of the domain Ω_a). In other words, here we have already factored in the periodicity with respect to k. An equivalent formulation in terms of quasi-momenta is as follows:

Theorem 3.4.6 *For any* $a \in (0, \infty]$, *the Floquet transform*

$$\mathcal{U}_\Gamma : f \mapsto \hat{f}(k, x)$$

is a topological isomorphism of the space $L^{2,a}(\mathbb{R}^n)$ *onto the subspace of* Γ^**-periodic functions in* $A(\mathcal{D}_a, L^2(\mathbf{W}))$.

3.4.2 Smooth Case

Dealing with differential operators, we will also be interested in the behavior of Sobolev spaces $H^s(\mathbb{R}^n)$ under the Floquet transform. We would like to obtain a Plancherel-type theorem, that is, to describe the image under Floquet transform of Sobolev space $H^s(\mathbb{R}^n)$.

This seems to be a piece of cake, just replacing everywhere L^2 with H^s. Well, not so fast. Repeating the proof of Theorem 3.4.1, one obtains the following statement:

Lemma 3.4.7 *The Floquet transform is an isometric mapping from $H^s(\mathbb{R}^n)$* ***into*** *the space $L^2(\mathbb{T}^*, H^s(\mathbf{W}))$. In particular, the following equality holds:*

$$\|f\|^2_{H^s(\mathbb{R}^n)} = \int_{\mathbb{T}^*} \|\hat{f}(z,\cdot)\|^2_{H^s(\mathbf{W})} d'z, \tag{3.4.7}$$

where $\mathbf{W}$ is the Wigner–Seitz cell of Γ and $d'z$ is the normalized Haar measure on $\mathbb{T}^*$.

It might seem that this lemma provides the needed analog of Plancherel's theorem, but this impression is misleading. What is missing then? The statement in fact **does not** claim that the Floquet transform is surjective (i.e., that its range is the whole space) between the spaces in question. As we will see now, its range on $H^s(\mathbb{R}^n)$ does **not** cover the whole space $L^2(\mathbb{T}^*, H^s(\mathbf{W}))$:

Proposition 3.4.8 *$\mathcal{U}_\Gamma$ is an isometry from $H^s(\mathbb{R}^n)$* ***onto a proper subspace*** *of $L^2(\mathbb{T}^*, H^s(W))$.*

Indeed, given a function $g \in L^2(\mathbb{T}^*, H^s(\mathbf{W}))$, one can recover using (3.2.10) its Fourier coefficients $g_\gamma \in H^s(\mathbf{W})$ and build out of them a function $f(x)$ such that for $x \in \mathbf{W} + \gamma$ one has $f(x) = g_\gamma(x - \gamma)$ a.e. The resulting function belongs to $L^2(\mathbb{R}^n)$, its restriction to $\mathbf{W} + \gamma$ belongs to $H^s(\mathbf{W} + \gamma)$ for any $\gamma \in \Gamma$, its Floquet transform is g, and finally

$$\sum_{\gamma\in\Gamma} \|f\|^2_{H^s(\mathbf{W}+\gamma)} < \infty.$$

This, however, does **not** imply that $f \in H^s(\mathbb{R}^n)$, for the obvious reason that smoothness is not enforced across the boundaries between the copies $\mathbf{W} + \gamma$ of the fundamental domain $\mathbf{W}$.

Clearly, some boundary conditions are needed to glue the pieces on each shifted fundamental domain $W + \gamma$ together. Indeed, these are easy to establish, for instance, when $s = 2$ (or any $s \in \mathbb{Z}^+$). The condition requires matching of the boundary Dirichlet and Neumann data for the function. This understanding also provides a way of fixing the problem and describing the range of the transform explicitly.

In the case of an integer $s > 0$ and a "nice" fundamental domain $\mathbf{W}$, one can provide an internal description of this subspace. Indeed, assume for simplicity that $\Gamma = \mathbb{Z}^n$ (the general lattice can be dealt with similarly) and $\mathbf{W} = (0,1)^n$ be the fundamental cube.

We introduce Definition 3.4.9.

Definition 3.4.9 For any $k \in \mathbb{C}^n$, the subspace

$$H_k^s(\mathbf{W}) \subset H^s(\mathbf{W}),$$

or just H_k^s, consists of restrictions to $\mathbf{W}$ of functions $f \in H_{\text{loc}}^s(\mathbb{R}^n)$ that satisfy the cyclic (Floquet) condition analogous to (3.2.6):

$$f(x+\gamma) = e^{ik\cdot\gamma} f(x) \text{ a.e.} \tag{3.4.8}$$

The following statement is natural and not hard to prove:

Lemma 3.4.10 *For $s \geq 1$ an integer let the quasi-momentum $k = (k_1, \dots, k_n)$ belong to $\mathbb{C}^n$. A function $f \in H^s(\mathbf{W})$ belongs to $H_k^s(\mathbf{W})$ if and only if the following boundary conditions are satisfied for any integers $0 \leq m \leq s-1$ and $1 \leq j \leq n$:*

$$\left.\frac{\partial^m f}{\partial x_j^m}\right|_{x_j=1} = e^{ik_j} \left.\frac{\partial^m f}{\partial x_j^m}\right|_{x_j=0}. \tag{3.4.9}$$

It is, however, easier to avoid this and to adopt a much more telling (and applicable for any, even fractional, $s \geq 0$) view. The clue on how to do this is derived from the automorphicity property (3.2.6)–(3.2.7).

It is straightforward to check that the following lemma holds.

Lemma 3.4.11 *The subspace $H_k^s = H_k^s(\mathbf{W}) \subset H^s(\mathbf{W})$ is closed.*

There is another useful way to describe the class of functions we are interested in. Let again $k = (k_1, \dots, k_n) \in \mathbb{C}^n$ be a quasi-momentum and $f \in H_k^s(\mathbf{W})$. We would like to do exactly what we have warned against: to interpret f as a "function" on the torus $\mathbb{T} = \mathbb{R}^n/\Gamma = \mathbf{W}/\Gamma$. The problem with this is that since the function is not periodic, after pushing it down to the torus, it becomes less smooth at the image of the former boundaries of $\mathbf{W}$. There is a simple way to take care of this, though.

One can define a smooth (in fact, analytic) complex line bundle $\mathbf{E_k}$ on $\mathbb{T}$ as follows. Let us fix $k \in \mathbb{C}^n$ and consider first the trivial line bundle

$$\mathbf{E} := \mathbb{R}^n \times \mathbb{C} \mapsto \mathbb{R}^n$$

over $\mathbb{R}^n$. We define a (k-dependent) Γ-action on E as follows:

$$(x, \lambda) \to (x+\gamma, e^{ik\cdot\gamma}\lambda), \qquad \gamma \in \Gamma. \tag{3.4.10}$$

Definition 3.4.12 We denote by $\mathbf{E_k}$ **the analytic complex line bundle over the torus** $\mathbb{T}$ obtained by factoring out the action τ_k of Γ on the bundle E.

Now the following claims are immediate to obtain:

Proposition 3.4.13 *Let $s \geq 0$. Then, for any k,*

1. *H^s_k is a closed subspace in $H^s(W)$.*
2. *$H^0_k = L_{2,k} = L^2(W)$.*
3. *$H^s_{k+k'} = H^s_k$ for any $k' \in \Gamma^*$.*
4. *A function u belongs to H^s_k if and only if*

$$\begin{aligned} u(x) &= e^{ik\cdot x}v(x), \\ &\text{with } v(x) \text{ being } \Gamma\text{-periodic, i.e.,} v \in H^s(\mathbb{T}). \end{aligned} \tag{3.4.11}$$

5. *Moreover, $\mathcal{H}^s := \bigcup_k H^s_k$ is an analytic Banach vector sub-bundle of the trivial bundle $\mathbb{C}^n \times H^s(W) \mapsto \mathbb{C}^n$ over $\mathbb{C}^n$.*
6. *There is a natural correspondence between Γ-automorphic functions on $\mathbb{R}^n$ with a quasi-momentum k and sections of the linear bundle E_k over $\mathbb{T}$.*
7. *Under this correspondence, elements of H^s_k correspond to H^s-sections of the bundle E_k.*
8. *"Any" Γ-periodic operator H preserves the Γ-automorphicity. In particular, a periodic elliptic operator H defines an elliptic operator $H(k)$ on sections of the bundle E_k.*[5]

In simpler terms, the subspace $H^s_k(\mathbf{W})$ "changes analytically" with the change of the quasi-momentum k. This means that there exists a holomorphic (entire) operator function $P(k)$ in $\mathbb{C}^n$, whose values are continuous projection operators in $H^s(\mathbf{W})$ and the range of $P(k)$ coincides with $H^s_k(\mathbf{W})$ (the fiber of the bundle $\mathcal{H}^s$) for any quasi-momentum $k \in \mathbb{C}^n$.

We can now formulate the exact range theorem for the Floquet transform in Sobolev spaces:

Theorem 3.4.14 *Let $s > 0$.*

- *$\mathcal{U}_\Gamma$ is isometry from $H^s(\mathbb{R}^n)$ onto the space $L^2(\mathbb{T}^*, \mathcal{H}^s)$ of the L^2-sections of the bundle $\mathcal{H}^s$.*
- *The section $\mathcal{U}_\Gamma f(\cdot, k)$ is infinitely differentiable on $\mathbb{T}^*$ if and only if the norms $\|f\|_{H^s(W+\gamma)}$ decay faster than any power of $\|\gamma\|$ when $\|\gamma\| \mapsto \infty$.*
- *The section $\mathcal{U}_\Gamma f(\cdot, k)$ is analytic in a complex neighborhood of $\mathbb{T}^*$ if and only if the norms $\|f\|_{H^s(W+\gamma)}$ decay exponentially fast when $\|\gamma\| \mapsto \infty$.*

[5] The detailed discussion of the action of Floquet transform on periodic operators is provided in the corresponding section later in this chapter, so at the moment the reader can skip this part of the Proposition.

One can notice that the bundle $\mathcal{H}^s$ over $\mathbb{C}^n$ is in a natural sense Γ^*-invariant, and, hence, in particular, it can be pushed down to the space of Floquet multipliers Ω (3.3.2). One of the ways to see this is to notice that the fiber $H^s_k(\mathbf{W})$ in fact depends only on k modulo the reciprocal lattice Γ^*, that is, it depends on the Floquet multiplier z only (and hence can be legitimately denoted by $H^s_z(\mathbf{W})$). Indeed, the action (3.4.10) of Γ on $\mathcal{B}$ can be written as $(x,\lambda) \to (x+\gamma, z^\gamma\lambda)$. One could also show that there exists a Γ^*-periodic holomorphic projector operator function $P(k)$ in $H^s(\mathbf{W})$ whose range is $H^s_k(\mathbf{W})$ (e.g., see Appendix A). Abusing notations, we will use the same notation $\mathcal{H}^s$ to denote the resulting bundle over Ω.

We are now able to establish analogs of Paley–Wiener theorems in the smooth case. In order to do so, let us introduce some auxiliary function spaces. The first type of these spaces will incorporate local finite smoothness with an exponential decay at infinity.

Definition 3.4.15

- The space $\mathcal{S}^s(\mathbb{R}^n)$ consist of all functions $f \in H^s_{\text{loc}}(\mathbb{R}^n)$ such that for any $N > 0$ the following expression is finite:

$$\rho_{s,N}(f) = \sup_{\gamma\in\Gamma} \|f\|_{H^s(\mathbf{W}+\gamma)}|\gamma|^N. \tag{3.4.12}$$

  The space $\mathcal{S}^s(\mathbb{R}^n)$ is equipped with the Fréchet topology defined by the semi-norms $\rho_{s,N}$.
- Let $a \in (0,\infty]$. The space $H^{s,a}(\mathbb{R}^n)$ consist of all functions $f \in H^s(\mathbb{R}^n)$ such that for any $0 < b < a$, the following expression is finite:

$$\psi_{s,b}(f) = \sup_{\gamma\in\Gamma} \|f\|_{H^s(\mathbf{W}+\gamma)}e^{b|\gamma|}. \tag{3.4.13}$$

  This space is equipped with the Fréchet topology defined by the semi-norms $\psi_{s,b}$.

The next result combines the Plancherel and Paley–Wiener type statements for the Floquet transform of H^s-functions. We use the Floquet multiplier z rather than quasi-momentum variables k here (it is a simple exercise to restate the following statements in terms of quasi-momenta).

Theorem 3.4.16 *Let $s \geq 0$ and $a \in (0,\infty]$. Then*

(1) *The Floquet transform is a topological isomorphism from the Fréchet space $\mathcal{S}^s(\mathbb{R}^n)$ to $C^\infty(\mathbb{T}^*,\mathcal{H}^s)$, where $C^\infty(\mathbb{T}^*,\mathcal{H}^s)$ is the space of C^∞ sections of the bundle $\mathcal{H}^s$ over $\mathbb{T}^*$.*
(2) *The Floquet transform is an isometric isomorphism from the space $H^s(\mathbb{R}^n)$ to $L^2(\mathbb{T}^*,\mathcal{H}^s)$, where $L^2(\mathbb{T}^*,\mathcal{H}^s)$ is the space of L^2-sections of the bundle $\mathcal{H}^s$ torus $\mathbb{T}^*$.*

(3) *Floquet transform is a topological isomorphism from the Fréchet space* $H^{s,a}(\mathbb{R}^n)$ *to* $A(\Omega_a, \mathcal{H}^s)$*, the space of analytic sections of the bundle* $\mathcal{H}^s$ *over the domain* Ω_a *(see (3.3.4)).*

This result can be reformulated in a somewhat more simple-minded fashion as follows:

Theorem 3.4.17 *Let* $s \geq 0$ *and* $a \in (0, \infty]$*. Then*

(1) *The Floquet transform is a topological isomorphism from the space* $\mathcal{S}^s(\mathbb{R}^n)$ *onto the subspace of the space* $C^\infty(\mathbb{T}^*, H^s(\mathbf{W}))$ *that consists of all vector functions* $g \in C^\infty(\mathbb{T}^*, H^s(\mathbf{W}))$ *such that* $g(z) \in H^s_z(\mathbf{W})$ *for all* $z \in \mathbb{T}^*$.
(2) *The Floquet transform is an isometric isomorphism from the space* $H^s(\mathbb{R}^n)$ *onto the subspace of the space* $L^2(\mathbb{T}^*, H^s(\mathbf{W}))$ *that consists of all vector functions* $g \in L^2(\mathbb{T}^*, H^s(\mathbf{W}))$ *such that* $g(z) \in H^s_z(\mathbf{W})$ *for a.e.* $z \in \mathbb{T}^*$.
(3) *The Floquet transform is a topological isomorphism from the Fréchet space* $H^{s,a}(\mathbb{R}^n)$ *onto the subspace of the space* $A(\Omega_a, H^s(\mathbf{W}))$ *that consists of all vector functions* $g \in A(\Omega_a, H^s(\mathbf{W}))$ *such that* $g(z) \in H^s_z(\mathbf{W})$ *for any* $z \in \Omega_a$*. Here* $A(\Omega_a, H^s(\mathbf{W}))$ *is the Fréchet space of analytic* $H^s(\mathbf{W})$*-valued functions in the domain* Ω_a.

The proofs of these theorems can be found in Kuchment (1993, Chapters 2 and 3).

A somewhat more special class of spaces and corresponding Paley–Wiener theorems are needed for some problems, for example, when studying in Section 6.3.2 "impurity"-embedded eigenvalues of perturbed operators (see also Kuchment, 1993, for more details and applications).

We now test the patience of the readers by introducing more function spaces.

Definition 3.4.18 Let H be a Hilbert space and $p > 1$. We define the space $A_p(\mathbb{C}^n, H)$ consisting of all entire H-valued $2\pi\mathbb{Z}^n$-periodic functions $\phi(k)$ in $\mathbb{C}^n$ that have order p and minimal type (e.g., Ronkin, 1974). This means that for any $a > 0$, the following expression is finite:

$$q_a(\phi) = \sup_{k \in \mathbb{C}^n} \|\phi(k)\| e^{-a|k|^p} < \infty. \tag{3.4.14}$$

The space $A_p(\mathbb{C}^n, H)$ is a Fréchet space with respect to the semi-norms q_a.

For any $2\pi\mathbb{Z}^n$-periodic holomorphic sub-bundle $\mathcal{E}$ of the trivial bundle $\mathbb{C}^n \times H$, we define by $A_p(\mathbb{C}^n, \mathcal{E})$ the subspace of $A_p(\mathbb{C}^n, H)$ consisting of functions ϕ which are also sections of $\mathcal{E}$, that is, for any point z, $\phi(z)$ belongs to the fiber $\mathcal{E}_z \subset H$ of $\mathcal{E}$ over z.

Definition 3.4.19 Let $s \geq 0$ and $p \in [1, \infty)$. We denote by $\Psi^s_{n,p}$ the following space of functions $f(x)$ on $\mathbb{R}^n$:

$$\left\{ f \in H^s_{\text{loc}}(\mathbb{R}^n) \mid d_b(f) := \sup_{\gamma \in \mathbb{Z}^n} \|f\|_{H^s(\mathbf{W}-\gamma)} e^{b|\gamma|^p} < \infty \text{ for all } b \geq 0 \right\}, \quad (3.4.15)$$

where, as before, **W** is a fundamental domain of the action of $\mathbb{Z}^n$ by shifts (i.e., the Wigner–Seitz cell).

The space $\Psi^s_{n,p}$ is a Fréchet space with respect to the semi-norms d_b.

We can now formulate the Paley–Wiener-type theorem for the space just introduced:

Theorem 3.4.20 *Let $s \geq 0$ and $p > 1$. The Floquet transform is an isomorphism of the space $\Psi^s_{n,p}$ onto the space $A_{p/(p-1)}(\mathbb{C}^n, \mathcal{E}^s)$.*

The proof, which is not very complicated, of this theorem can be found in Kuchment (1993, Theorem 2.2.2).

3.4.3 Orthogonality of Shifts: Preparing for Wannier Functions

In this short section, we formulate a simple conclusion, which will be important for considerations of the so-called **Wannier functions** in Section 7.2.

We will be interested here for the question of when all lattice shifts of an L^2-function $f(x)$ are mutually orthogonal.

Theorem 3.4.21 *For $f \in L^2(\mathbb{R}^n)$, the functions $f(x-\gamma)$ are mutually orthogonal for all $\gamma \in \Gamma$ if and only if the Floquet transform $\hat{f}(k, \cdot)$ of f has a constant L^2 norm on the Wigner–Seitz cell W:*

$$\|\hat{f}(k, \cdot)\|_{L^2(W)} = \text{constant}. \quad (3.4.16)$$

Proof The proof is rather straightforward. Indeed, according to (3.2.6), the γ-shift of f results in the multiplication of $\hat{f}(k, \cdot)$ by $\exp(ik \cdot \gamma)$. As a result of Theorem 3.4.1, it is sufficient to check the mutual orthogonality of the functions $\exp(ik \cdot \gamma)\hat{f}(k, \cdot)$ in $L^2(\mathbb{T}^*, L^2(W))$. This results in verifying the equality

$$\int_{\mathbb{T}^*} \exp(ik \cdot \gamma) \|\hat{f}(k, \cdot)\|^2_{L^2(W)} d'k = 0$$

for all $\gamma \in \Gamma, \gamma \neq 0$. In other words, the scalar function $\|\hat{f}(k, \cdot)\|^2_{L^2(W)}$ on $\mathbb{T}^*$ must have all Fourier coefficients, except the constant term, equal to zero. This happens if and only if this function is a constant. □

3.5 Modified Floquet Transform

We see that when using the Floquet transform, one has to work with the restrictions of a fixed differential operator to the "rotating" subspaces $H^s_k(\mathbf{W})$ in $H^s(\mathbf{W})$. To put it differently, one needs to deal with boundary conditions depending on the quasi-momentum k (or the Floquet multiplier z). In some cases, it is more convenient to have constant boundary conditions, paying for this the price of dealing with a differential expression depending on k. This can be achieved using a modification of the Floquet transform.

Let us return to the cyclic property (3.2.6) of Floquet transform $\hat{f}(k,x) = \mathcal{U}_\Gamma f(k,x)$ of a function $f(x)$:

$$\hat{f}(k, x+\gamma) = e^{ik\cdot\gamma}\hat{f}(k,x).$$

This condition depends on k and thus causes "rotation" of the space H^s_k. However, this can be easily rectified. Indeed, the function $e^{ik\cdot x}$ satisfies exactly the same property with respect to x. Thus, dividing $\hat{f}$ by $e^{ik\cdot x}$, we make $\hat{f}$ periodic with respect to x. We formalize this introducing the following **modified Floquet transform**:

$$\mathcal{V}_\Gamma f(k,x) := e^{-ik\cdot x}\mathcal{U}_\Gamma f(k,x) = e^{-ik\cdot x}\hat{f}(k,x). \tag{3.5.1}$$

The resulting function $\mathcal{V}_\Gamma f$ satisfies the corresponding modified cyclicity conditions with respect to k and x:

$$\mathcal{V}_\Gamma f(k+\kappa, x) = e^{-i\kappa\cdot x}\mathcal{V}_\Gamma f(k,x), \qquad \kappa \in \Gamma^* \tag{3.5.2}$$

$$\mathcal{V}_\Gamma f(k, x+\gamma) = \mathcal{V}_\Gamma f(k,x), \qquad \gamma \in \Gamma. \tag{3.5.3}$$

We see that with respect to x, this function is Γ-periodic and, hence, can be interpreted **for any value of** k as a function on $\mathbb{T}$. The price to pay is the disappearance of periodicity with respect to k, which is replaced now with the cyclicity condition (3.5.2).

Plancherel and Paley–Wiener-type theorems can be reformulated in terms of the modified Floquet transform $\mathcal{V}_\Gamma$. For instance, the Plancherel-type statement is explained in the next theorem:

Theorem 3.5.1 *For any $s \geq 0$, the modified Floquet transform $\mathcal{V}_\Gamma$ is an isometric isomorphism of the space $H^s(\mathbb{R}^n)$ onto $L^2(\mathbf{B}, H^s(\mathbb{T}))$. Equivalently, it provides the following direct integral expansion*

$$H^s(\mathbb{R}^n) = \int_{\mathbf{B}}^{\oplus} H^s(\mathbf{W})d'k. \tag{3.5.4}$$

Notice that the distinction between expansions (3.4.4) and (3.5.4) is that in the second case, due to the loss of periodicity with respect to k, integration is done over Brillouin zone $\mathbf{B}$ rather than the torus $\mathbb{T}^*$. On the other hand, the fibers of the direct integral expansion are now all equal to a fixed space $H^s(\mathbb{T})$ rather than being dependent on the quasi-momentum k (as the spaces $H^s_k(\mathbf{W})$ are).

One can also obtain Paley–Wiener-type theorems for $\mathcal{V}_\Gamma$ (see Section 3.9), which we will not discuss in detail now.

When using the modified Floquet transform, it is better to stick to quasi-momenta rather than Floquet multipliers, due to the loss of periodicity with respect to k.

3.6 Floquet Transforms of Periodic Operators

While we have studied so far the action of the Floquet transform on function spaces, now it is time to understand what it does to periodic differential operators.

Consider a linear partial differential expression of order m

$$L(x, D) = \sum_{|\alpha|\leq m} a_\alpha(x) D^\alpha \tag{3.6.1}$$

in $\mathbb{R}^n$, where $D = \frac{1}{i}\frac{\partial}{\partial x}$, $\alpha = (\alpha_1, \ldots, \alpha) \in (\mathbb{Z}^+)^n$ is a multi-index, $|\alpha| = \sum \alpha_j$, and $D^\alpha = i^{-|\alpha|} \dfrac{\partial^{|\alpha|}}{\partial^{\alpha_1} x_1 \cdots \partial^{\alpha_n} x_n}$. According to the approach adopted in this book, we assume that the coefficients a_α of the operator belong to $C^\infty(\mathbb{R}^n)$ and are *periodic* with respect to a lattice Γ in $\mathbb{R}^n$, that is, that

$$a_\alpha(x + \gamma) = a_\alpha(x) \tag{3.6.2}$$

for all α and $x \in \mathbb{R}^n$, $\gamma \in \Gamma$. As it was mentioned in the Introduction, the infinite (and even lesser) differentiability assumption on the coefficients is a major overkill and is essentially never needed in its full strength (or even close to it). However, it simplifies considerations a lot without changing the main ideas of the theory. A reader interested in relaxed or minimal requirements for the coefficients is referred to the comments sections of the chapters, where additional comments and references to the literature are provided.

We will need to use the notions of the **symbol**

$$L(x, \xi) = \sum_{|\alpha|\leq m} a_\alpha(x) \xi^\alpha \tag{3.6.3}$$

and **principal symbol**

$$L_0(x, \xi) = \sum_{|\alpha|=m} a_\alpha(x) \xi^\alpha \tag{3.6.4}$$

of the operator L.

For instance, for the Schrödinger operator $H = -\Delta + V(x)$, the symbol is $H(x, \xi) = \xi^2 + V(x)$ and the principal symbol is $H_0(x, \xi) = \xi^2$.

Although most of the considerations of this chapter work for any such operator, in the rest of the book we will need to assume that the operator is **elliptic**.[6] We adopt the following standard definition of ellipticity:

Definition 3.6.1 Operator L is **elliptic**, if its principal symbol $L_0(x, \xi)$ does not vanish for any $x \in \mathbb{R}^n$ and $\xi \neq 0$ in $\mathbb{R}^n$.

We will often need to consider the operator $L^t(x, D)$ that is **dual** (or **transposed**) to the operator $L(x, D)$. The duality is understood with respect to the following **bilinear** form on $L^2(\mathbb{R}^n)$:

$$(f, g) = \int_{\mathbb{R}^n} f(x)g(x)\, dx. \tag{3.6.5}$$

In other words, for sufficiently smooth and compactly supported functions, one has

$$(Lf, g) = (f, L^t g). \tag{3.6.6}$$

Remark 3.6.2

(1) If L is elliptic, it is easy to see that L^t is elliptic as well.

(2) The reader must notice the bilinearity of the form (3.6.5) instead of the more common sesquilinearity. In other words, we do not use complex conjugate of the function f or g in (3.6.5). This will be more convenient for our purpose.

Using the **sesquilinear duality**

$$\langle f, g \rangle = \int_{\mathbb{R}^n} f(x)\overline{g(x)}\, dx, \tag{3.6.7}$$

instead, one arrives to the notion of the **adjoint** operator L^*.

Consider a function $u \in H^m(\mathbb{R}^n)$. Then function Lu is naturally defined and, due to our conditions on the coefficients, it belongs to $L^2(\mathbb{R}^n)$. In other words, we have a (continuous) linear operator $L : H^m(\mathbb{R}^n) \mapsto L^2(\mathbb{R}^n)$. According to our study of Floquet transform, both expressions $\mathcal{U}_\Gamma u$ and $\mathcal{U}_\Gamma(Lu)$ are defined for $u \in H^m(\mathbb{R}^n)$. Since the coefficients in (3.6.1) are Γ-periodic, operator L commutes with the action of Γ by shifts. We hence expect that Floquet transform will (at least, partially) **diagonalize** L. In other words, in the direct integral expansion (3.4.4), the operator L must look like multiplication by an operator-valued function $L(k)$.

Let us do a short formal calculation to confirm this and to find the operator function we have just mentioned. Taking into account the definition (3.2.4) of

[6] Sometimes, for example, when considering time periodic parabolic equations, one can use **hypoellipticity** instead. This will be addressed in Kuchment (2027); see also Kuchment (1982, 1993).

Floquet transform and invariance of L with respect to Γ-shifts, we get

$$\begin{aligned}\mathcal{U}_\Gamma(Lu)(k,x) &= \sum_{\gamma\in\Gamma}(Lu)(x+\gamma)e^{-ik\cdot\gamma} = \sum_{\gamma\in\Gamma}L(x,D_x)(u(x+\gamma))e^{-ik\cdot\gamma}\\ &= L(x,D_x)(\mathcal{U}_\Gamma u)(k,x).\end{aligned} \tag{3.6.8}$$

Here, we have emphasized, through the notation $D_x = \frac{1}{i}\partial_x$, that in the last two expressions the differential operator L acts with respect to the x-variable.

At first blush, (3.6.8) might create the wrong impression that in the right-hand side the same operator L acts for any value of the quasi-momentum k. This would mean that in the direct integral expansion, for every k, the same operator acts on the fiber above k. This is certainly not true. The differential expression $L(x, D_x)$ truly does not depend on k, however the boundary conditions, and thus the **domain** of the operator, do change with k. Let us recall (see, for instance, Theorem 3.4.17) that in the direct integral expansion of the space $H^m(\mathbb{R}^n)$, the fiber over k is the space $H^m_k = H^m_k(\mathbf{W})$ (or $H^m_z(\mathbf{W})$, if we prefer to deal with Floquet multipliers z rather than quasi-momenta k). Elements of this space are H^m-functions on $\mathbf{W}$ that satisfy k-dependent boundary conditions (3.4.9). Thus, if we define the operator $L(k)$ by the differential expression $L(x, D_x)$ on the space H^m_k, we get

$$\mathcal{U}_\Gamma(Lu)(k,\cdot) = L(k)(\mathcal{U}_\Gamma u)(k,\cdot). \tag{3.6.9}$$

Another way to put it is that $\mathcal{U}_\Gamma L \mathcal{U}_\Gamma^{-1}$ is the operator of multiplication by the operator-valued function $L(k)$. One can also write this as the direct integral decomposition:

$$L = \int_{\mathbf{B}}^{\oplus} L(k)\, dk = \int_{\mathbb{T}^*}^{\oplus} L(z)\, dz. \tag{3.6.10}$$

So, as we expected, Floquet transform partially diagonalizes periodic operators. This is what makes it so useful for studying periodic PDEs, the same way Fourier transform is handy when studying constant coefficient linear PDEs. In the constant coefficient case, after Fourier transform a constant coefficient operator $P(D)$ becomes multiplication operator by its scalar (polynomial) symbol $P(\xi)$. In comparison, after Floquet transform, a periodic operator $L(x, D)$ becomes multiplication operator by the polynomial operator function $L(k)$. One sees that the operator function $L(k)$ plays the role of an operator-valued symbol of L.

It is well understood that behavior of the symbol $P(\xi)$ with respect to ξ plays the crucial role in determining properties of the constant coefficient operator $P(D)$ (e.g., Ehrenpreis, 1970; Hörmander, 2005; Palamodov, 1970). Hence, one expects that it would be also useful to know the behavior of $L(k)$ with respect to k.

So far we have described $L(k)$ as follows: It is the restriction of a fixed operator $L(x, D)$ acting from $H^m(\mathbf{W})$ to $L^2(\mathbf{W})$ to an analytically "rotating" subspace

$H_k^m(\mathbf{W}) \subset H^m(\mathbf{W})$. Well, this description of $L(k)$ is not too revealing. It would be nice to have instead an operator that has a fixed domain, while its differential expression could depend on k in some nice manner. In terms of vector bundles, we would like to **trivialize** the analytic vector bundle $\mathcal{H}^m$. As it happens, the modified Floquet transform does exactly this. Indeed, a direct calculation establishes the following auxiliary result:

Lemma 3.6.3

(1) *The modified Floquet transform* $\mathcal{V}_\Gamma$ *reduces the operator* L *to the operator of multiplication by the operator-valued polynomial*

$$L(k) := L(x, D_x + k) = \sum_\beta \frac{1}{\beta!}\frac{\partial^{|\beta|} L}{\partial \xi^\beta}(x, D_x)k^\beta \tag{3.6.11}$$

acting from $H^m(\mathbb{T})$ *to* $L^2(\mathbb{T})$.

(2) *The operators* $L(k + k_0)$ *and* $L(k)$*, for* $k_0 \in \Gamma^*$*, are unitarily equivalent, with equivalence provided by multiplication by* $e^{ik_0\cdot x} \in C^\infty(\mathbb{T})$*:*

$$L(k + k_0) = e^{-ik_0\cdot x} L(k) e^{ik_0\cdot x} \tag{3.6.12}$$

(3) *If the operator* $L(x, D)$ *is elliptic, then all operators* $L(k) : H^m(\mathbb{T}) \to L^2(\mathbb{T})$ *are Fredholm with Fredholm index equal to zero.*[7]

Before proving this statement, let us first explain some of the notation used. Let

$$L(x, \xi) = \sum_{|\alpha| \le m} a_\alpha(x)\xi^\alpha$$

be the (polynomial in ξ) symbol of $L(x, D)$. Then one can take its derivatives $\frac{\partial^{|\beta|} L}{\partial \xi^\beta}$ with respect to ξ and use them as symbols of new differential operators, which are exactly shown in (3.6.11).

Proof As we discussed in Section 3.5, in order to move to the modified Floquet transform from the regular one, the functions from $H_k^m(\mathbf{W})$ are divided by the function $e^{ik\cdot x}$. Hence, the differential expression of the operator $L(k)$ after the modified transform becomes $e^{-ik\cdot x} L(x, D) e^{ik\cdot x} = L(x, D + k)$. This proves the first two statements of the lemma. The third statement is the standard fact of the theory of elliptic operators on compact manifolds (see, e.g., Taylor, 2011). □

[7] A Fredholm operator has closed range and finite dimensional kernel and co-kernel. Its index is the difference of the dimensions of its kernel and co-kernel (see Appendix B or Kato, 1995; Gohberg and Kreĭn, 1957; Krein, 1982 for details). The index of elliptic operators in nontrivial bundles over the torus might be nonzero (see, e.g., Atiyah and Segal, 1968; Atiyah and Singer, 1968a, 1968b; Lawson and Michelsohn, 1989).

Example 3.6.4 Consider the Schrödinger operator

$$H = -\Delta + V(x) = D^2 + V(x)$$

with a Γ-periodic potential $V(x)$. Then

$$\begin{aligned} H(k)(x,D) &:= H(x,D+k) = (D+k)^2 + V(x) \\ &= D^2 + 2k \cdot D + k^2 + V(x) = -\Delta - 2i \sum_j k_j \frac{\partial}{\partial x_j} + \sum k_j^2 + V(x) \\ &= k^2 + 2k \cdot D + (-\Delta + V(x)). \end{aligned}$$

Lemma 3.6.3 achieves exactly what we desired: Now the domain $H^m(\mathbb{T})$ of $L(k)$ is constant and consists of periodic H^m-functions on the torus, while the differential expression for $L(k)$ depends on k in a polynomial manner. Moreover, its all dependence on k is contained in lower (differential) order terms. Also, the coefficient at k^α in (3.6.11) is a differential operator of order $m - |\alpha|$, so its order drops with increase in $|\alpha|$. As we will see, all these simple conclusions are essential.

One might ask, what is the advantage of reducing a single periodic differential operator $L(x, D)$ to a whole family of operators $L(k)$, each of which is a partial differential operator again. Don't we make our study more complicated? The main advantage of this diagonalization is that each of the operators $L(k)$ acts on the **compact** manifold $\mathbb{T}$, while the original operator L acts on the whole non-compact $\mathbb{R}^n$. The reader can foresee that this will make the whole world of a difference, in particular for spectral theory, as well as for other problems. For instance, compactness of $\mathbb{T}$ will imply that the operators $L(k)$ are Fredholm (see Appendix B) and have compact resolvents and thus **discrete spectra**, which is not true for the original operator L.

Lemma 3.6.5 *If the operator L is elliptic, then for any $k \in \mathbb{C}^n$ the operator $L(k) : H^{s+m}(\mathbb{T}) \mapsto H^s(\mathbb{T})$ is Fredholm, that is, is bounded with closed range and finite-dimensional kernel and co-kernel.*

This follows from the well-known Fredholm property of elliptic operators on compact manifolds (such as $\mathbb{T}$) (e.g., Theorem 2.40 in Egorov and Shubin, 1998; Lions and Magenes, 1972, or just about any source on elliptic operators.) This statement strengthens (3) in Lemma 3.6.3.

3.7 Another Look at Floquet–Bloch Functions

Let us recall that for a given lattice Γ a function $f(x)$ is a **Bloch function with the quasi-momentum** $k \in \mathbb{C}^n$, if it is represented as

$$f(x) = e^{ik\cdot x} p(x) \tag{3.7.1}$$

with a Γ-periodic function $p(x)$.

Remark 3.7.1 Bloch functions are often called Floquet functions or Bloch–Floquet functions. In this book, we reserve the name Floquet–Bloch function for a more general class of functions (see Definition 3.7.2).

We came to this definition considering the (discrete) shift-invariance with respect to a lattice Γ. Thus, such functions are analogs of exponential functions $e^{\xi \cdot x}$ playing major role in studying constant coefficient linear PDEs.

After Fourier transformation, exponential functions become delta-functions of the dual variable ξ. More generally, the **exponential polynomials** $e^{\xi \cdot x} \sum c_\alpha x^\alpha$ become under Fourier transform linear combinations of delta function and its derivatives at the point ξ.

Analogs of the exponential polynomials for the periodic case are the **Floquet–Bloch functions** with the quasi-momentum k:

Definition 3.7.2 A function of the form

$$f(x) = e^{ik\cdot x} \sum_{|\alpha| \leq N} x^\alpha q_\alpha(x), \tag{3.7.2}$$

where functions $q_\alpha(x)$ are Γ-periodic, is said to be a **Floquet–Bloch function** with the quasi-momentum k.

The natural question is whether the Floquet–Bloch function also correspond under Floquet transform to some kind of "delta function" of the quasi-momentum k. One expects that maybe after Floquet transform such functions become linear combinations of "delta functions" and their derivatives at the point k.

The goal of this section is to show that this is indeed the case, which turns out important for a variety problems concerning periodic operators of mathematical physics.

Remark 3.7.3 As we have mentioned earlier, in terms of group representation theory, the exponentials and Bloch functions are derived from irreducible representations of the relevant (abelian) group of symmetries. The exponential polynomials and Floquet functions arise from the so-called **indecomposable representations** (Jordan cells in the matrix theory).

3.7.1 Bloch Functions as "Delta-Functions"

Let us try first to think what could be called a "delta function" in k. Well, Floquet transform produces vector-valued, with values in $L^2(\mathbf{W})$ (rather than scalar) functions of k. Thus, in order to make a "delta-type" functional truly a functional rather than an operator, one needs to choose a function $q(x) \in L^2(\mathbf{W})$ on $\mathbf{W}$ and then to define a functional as follows:

$$f(x) \mapsto \int_{\mathbf{W}} \hat{f}(k_0, x) q(x) dx. \tag{3.7.3}$$

This would be a "delta function supported at k_0" with a factor $q(x)$ (which replaces a constant factor in case of exponents and Fourier transform).

Remark 3.7.4 One should notice that this calculation so far has been formal and cannot be done when, for instance, $f \in L^2(\mathbb{R}^n)$, since we know that the function $\hat{f}$ then will be just an L^2-function with respect to k and thus could not be evaluated at a point. For example, the additional assumption that

$$\sum_{\gamma \in \Gamma} \|f\|_{L^2(\mathbf{W}+\gamma)} < \infty$$

would do, since one can easily check that then $\hat{f}(k)$ will be **continuous** as a function with values in $L^2(\mathbf{W})$.

Now it is interesting to see what this functional corresponds to in the original x-space. Let us agree first that $q(x)$ (defined originally on the Wigner–Seitz cell $\mathbf{W}$ only) is extended Γ-periodically to the whole $\mathbb{R}^n$.

Theorem 3.7.5 *Bloch functions with quasi-momentum k_0 (and only them) correspond to functionals* (3.7.3).

Proof Involving the definition (3.2.4) and using periodicity of $q(x)$, we have

$$\begin{aligned}\int_{\mathbf{W}} \hat{f}(k_0, x) q(x) dx &= \sum_{\gamma \in \Gamma} e^{-ik_0 \cdot \gamma} \int_{\mathbf{W}} f(x+\gamma) q(x) dx \\ &= \int_{\mathbb{R}^n} f(x) \left(\sum_{\gamma \in \Gamma} e^{-ik_0 \cdot \gamma} \chi_{\mathbf{W}+\gamma}(x) q(x) \right) dx \\ &= \int_{\mathbb{R}^n} f(x) b(x)\, dx,\end{aligned} \tag{3.7.4}$$

where $\chi_U(x)$ is the characteristic function of a set U and

$$b(x) = \sum_{\gamma \in \Gamma} e^{-ik_0 \cdot \gamma} \chi_{\mathbf{W}+\gamma}(x) q(x).$$

It is easy to check now that $b(x)$ is in fact a Bloch function with quasi-momentum

$-k_0$. Indeed, one has $\chi_U(x+\gamma) = \chi_{U-\gamma}(x)$ and thus

$$\begin{aligned} b(x+\gamma_1) &= \sum_{\gamma\in\Gamma} e^{-ik_0\cdot\gamma}\chi_{\mathbf{W}+\gamma}(x+\gamma_1)q(x) \\ &= \sum_{\gamma\in\Gamma} e^{-ik_0\cdot((\gamma-\gamma_1)+\gamma_1)}\chi_{\mathbf{W}+(\gamma-\gamma_1)}(x)q(x) \\ &= e^{-ik_0\cdot\gamma_1}\sum_{\gamma\in\Gamma} e^{-ik_0\cdot\gamma}\chi_{\mathbf{W}+\gamma}(x)q(x) \\ &= e^{-ik_0\cdot\gamma_1}b(x). \end{aligned}$$

So, our conclusion is that Bloch functions exactly correspond to "delta-type" functionals after Floquet transform, and the supports of such functionals provide the values of quasi-momenta. □

This result, being important by itself, will also help us now to define analogs of exponential polynomials.

Indeed, Fourier transforms of exponential polynomials are known to be linear combinations of a delta function and several of its derivatives. It is natural to look for functionals that after Floquet transform look similar (with the same caveat concerning vector values as discussed earlier). So, we want to have a functional that acts on a function $f(x)$ from an appropriate class (e.g., from $L_{2,\text{comp}}(\mathbb{R}^n)$) as follows:

$$f(x) \mapsto \sum_{|\alpha|\leq N}\int_{\mathbf{W}} \frac{\partial^{|\alpha|}\hat{f}}{\partial k^{\alpha}}|_{(k_0,x)} q_{\alpha}(x)\, dx\,. \tag{3.7.5}$$

Theorem 3.7.6 *The functional* (3.7.5) *can be written as*

$$f(x) \mapsto \int_{\mathbb{R}^n} f(x)b(x)\, dx,$$

where the function $b(x)$ can be represented as a "periodic exponential polynomial"

$$b(x) = e^{-ik_0\cdot x}\sum_{|\alpha|\leq N} x^{\alpha}p_{\alpha}(x), \tag{3.7.6}$$

the functions $p_j(x)$ are Γ-periodic, and the integer N is the same as in (3.7.5).

The converse statement is also correct: Functions of the type described in (3.7.6) *correspond to the functionals* (3.7.5).

Proof The proof goes by a direct calculation. It suffices to show here that a functional (3.7.5) with a single term in the sum can be represented as (3.7.6). So, let us have a functional

$$f(x) \mapsto \int\limits_{\mathbf{W}} \frac{\partial^{|\alpha|}\hat{f}}{\partial k^{\alpha}}(k_0,x)q(x)dx$$

with a Γ-periodic function $q(x)$. Referring again to the definition (3.2.4) of Floquet transform and using periodicity of $p(x)$, we obtain:

$$\begin{aligned}\int_{\mathbf{W}} \frac{\partial^{|\alpha|} \hat{f}}{\partial k^{\alpha}}(k_0, x) q(x)\, dx &= \sum_{\gamma \in \Gamma} (-i\gamma)^{\alpha} e^{-ik_0 \cdot \gamma} \int_{\mathbf{W}} f(x+\gamma) q(x)\, dx \\ &= \int_{\mathbb{R}^n} f(x) \left(\sum_{\gamma \in \Gamma} (-i\gamma)^{\alpha} e^{-ik_0 \cdot \gamma} \chi_{\mathbf{W}+\gamma}(x) q(x) \right) dx \\ &= \int_{\mathbb{R}^n} f(x) b(x)\, dx, \end{aligned} \tag{3.7.7}$$

where

$$b(x) = \sum_{\gamma \in \Gamma} (-i\gamma)^{\alpha} e^{-ik_0 \cdot \gamma} \chi_{\mathbf{W}+\gamma}(x) q(x).$$

The lemma will follow if we show that $b(x)$ can be represented in the form (3.7.6). In order to do so, we rewrite the formula for $b(x)$ as follows:

$$b(x) = (-i)^{|\alpha|} e^{-ik_0 \cdot x} \sum_{\gamma \in \Gamma} \gamma^{\alpha} e^{-ik_0 \cdot (\gamma - x)} \chi_{\mathbf{W}+\gamma}(x) q(x) = (-i)^{|\alpha|} e^{-ik_0 \cdot x} h(x).$$

So, it remains now to show that the sum $h(x)$ in this expression can be represented as

$$\sum_{\beta \leq \alpha} x^{\beta} q_{\beta}(x)$$

with Γ-periodic functions q_{β}.

Let us make first the following remark:

Remark 3.7.7 For arbitrary function $e(x)$ and any Γ-periodic function $q(x)$, the function

$$\sum_{\gamma \in \Gamma} e(\gamma - x) \chi_{\mathbf{W}+\gamma}(x) q(x)$$

is Γ-periodic. Indeed, let first $x \in \mathbf{W}$, then the sum reduces to $e(-x)q(x)$. If we now plug into the sum $x+\gamma$, it will reduce to $e(\gamma - (x+\gamma))q(x+\gamma)$, which is again equal to $e(-x)q(x)$. Thus, the sum represents a Γ-periodic function.

Taking into account this remark for $e(x) = e^{-ik_0 \cdot x}$, one can simplify $h(x)$ to

$$h(x) = \sum_{\gamma \in \Gamma} \gamma^{\alpha} \chi_{\mathbf{W}+\gamma}(x) q(x) = q(x) \sum_{\gamma \in \Gamma} \gamma^{\alpha} \chi_{\mathbf{W}+\gamma}(x)$$

with a different, but still Γ-periodic function $q(x)$. The periodic factor $q(x)$ in front of the sum is irrelevant for our purpose, so we can assume that

$$h(x) = \sum_{\gamma\in\Gamma} \gamma^\alpha \chi_{\mathbf{W}+\gamma}(x).$$

Representing γ^α as $(x + (\gamma - x))^\alpha$ and using binomial formula, we conclude that

$$h(x) = x^\alpha + \sum_{\beta<\alpha} c_\beta x^\beta \sum_{\gamma\in\Gamma} (\gamma - x)^{\alpha-\beta} \chi_{\mathbf{W}+\gamma}(x).$$

This, together with Remark 3.7.7, reduces $h(x)$ to the required form

$$h(x) = x^\alpha + \sum_{\beta<\alpha} x^\beta q_\beta(x)$$

with periodic functions $q_\beta(x)$. **This proves that any functional (3.7.5) corresponds to a Floquet–Bloch function with the quasi-momentum** $-k_0$. The converse statement is proven similarly, which finishes the proof. □

Now we have Bloch functions in our disposal, which are analogs of exponents and Floquet–Bloch functions, which are analogs of exponential polynomials. We will see that they are of extreme importance in the whole theory.

Another useful set of functions are the so-called **Wannier functions**, considerations of which we discuss in Chapter 7.

3.7.2 Floquet–Bloch Functions and Difference Operators

We now address, following Kha and Lin (2019), an alternative way of defining Bloch and Floquet functions (solutions), which often happens to be useful.

For any $\gamma \in \Gamma$, let us define the **first difference operator** Δ_γ acting on functions on $\mathbb{R}^n$ as follows:

$$\Delta_\gamma u(x) = u(\gamma x) - u(x). \tag{3.7.8}$$

The proof of the following lemma is straightforward:

Lemma 3.7.8 *A function u is Γ-periodic (i.e., a Bloch function with zero quasi-momentum) if and only if it is annihilated by Δ_γ for any $\gamma \in \Gamma$.*

In fact, it is sufficient to check this property for any set $\{\gamma_j\}$ of generators of Γ.

One wonders whether one can check in a similar way whether a function is a Bloch function with a nonzero quasi-momentum and whether Floquet functions allow for similar tests. In order to get the answer, we need to introduce a **twisted version of the first difference** that depends of the quasi-momentum:

$$\Delta_{\gamma;k} u(x) = \mathrm{e}^{-ik\cdot\gamma} u(\gamma x) - u(x). \tag{3.7.9}$$

We also introduce **iterated finite differences** of order N with quasi-momentum k as follows:

$$\Delta_{\gamma_1,\ldots,\gamma_N;k} = \Delta_{\gamma_1;k} \cdots \Delta_{\gamma_N;k}, \tag{3.7.10}$$

where $\gamma_j \in \Gamma$ (it will always be sufficient to use only elements, maybe repeated, of a fixed set of generators of Γ).

We can now answer the question by proving the following theorem:

Theorem 3.7.9 *A function $u(x)$ on $\mathbb{R}^n$ is a Floquet function of order N with quasi-momentum k if and only if it is annihilated by any iterated difference of order $N+1$ with quasi-momentum k:*

$$\Delta_{\gamma_1,\ldots,\gamma_{N+1};k} u = 0 \qquad \text{for all} \quad \gamma_1,\ldots,\gamma_{N+1} \in \Gamma,$$

(choosing γ_j from a fixed set of generators is sufficient). In particular, it is a Floquet function of order N with quasi-momentum 0 if and only if

$$\Delta_{\gamma_1} \cdots \Delta_{\gamma_{N+1}} u = 0 \qquad \text{for all} \quad \gamma_1,\ldots,\gamma_{N+1} \in \Gamma. \tag{3.7.11}$$

Proof Let us provide the proof for the case $k = 0$ first. As it has already been mentioned, the necessity and sufficiency of the condition (3.7.11) checks out easily for $N = 0$, where it boils down to $\Delta_\gamma u = 0$ for all $\gamma \in \Gamma$, that is, to the periodicity of u. Necessity for any N now follows easily by induction if one takes into account the representation (3.7.2). Indeed, one computes that (using the same standard basis $\{\gamma_j\}$ of $\mathbb{Z}^n$ as before)

$$\Delta_{\gamma_j} x^{(l_1,\ldots,l_n)} = l_j x^{(l_1,\ldots,l_j-1,\ldots,l_n)} + \text{ lower-order terms.} \tag{3.7.12}$$

Here "lower order terms" contain linear combinations of x^ls of strictly lower total degrees. Since the difference operators do not alter periodicity, we obtain that any Δ_{γ_j} reduces Floquet functions of order $N+1$ to the ones of order N, which concludes the induction step of the proof of necessity for $k = 0$.

Let us prove sufficiency, which also follows by induction with respect to the order of the Floquet function. We have already checked it for $N = 0$. Assume that this has been proven for orders up to N. Suppose that u satisfies $\Delta_{\gamma_1} \cdots \Delta_{\gamma_{N+2}} u = 0$ for any $\gamma_1,\ldots,\gamma_{N+2} \in \Gamma$. Take a set of generators $\{\gamma_j\}_{j=1}^n$ in Γ. Then, the functions $f_j := \Delta_{\gamma_j} u$ satisfy the condition of the Lemma for the order N. According to the induction hypothesis, we conclude that

$$f_j(x) = \left(\sum_{\substack{l = (l_1,\ldots,l_n) \in \mathbb{Z}_+^n \\ |l| \le N}} x^l p_{l,j}(x) \right)$$

with periodic functions $p_{l,j}$.

We claim that for any Floquet function f of order N (with $k = 0$) there exists a Floquet function v of order $N + 1$ such that $\Delta_{\gamma_j} v = f$. Without loss of generality, we assume that $j = 1$. Since the difference operator Δ_{γ_1} does not change periodic coefficients, it is sufficient to check the statement for $f = x^l$, where $|l| \leq N$. Now induction with respect to l_1 finishes the job. Namely, (3.7.12) provides a triangular linear system of equations for recursively determining the coefficients of v such that $\Delta_{\gamma_j} v = f$.

Therefore, a Floquet function v_1 of order $N+1$ exists such that $\Delta_{\gamma_1} v_1 = \Delta_{\gamma_1} u$. This means that the function $u_2 = u - v_1$ is periodic with respect to the one-parameter subgroup generated by γ_1. Since the condition of the Lemma is still satisfied for u_2, and since u_2 is γ_1-periodic, we can continue this process and find a Floquet function v_2 of order $N + 1$ that is γ_1-periodic and such that $u_3 = u - v_1 - v_2$ is periodic with respect to both γ_1 and γ_2. Continuing this process, we get the conclusion of the lemma for $k = 0$.

Let us now prove the statement for an arbitrary quasi-momentum k. According to the definition (3.7.2) of Floquet functions, any Floquet function u of order N with quasi-momentum k has the form $u = e^{ik\cdot x} v$, where v is a Floquet function of order N with quasi-momentum $k = 0$. A straightforward calculation shows that if any two functions u and v satisfy $u = e^{ik\cdot x} v$, then one has $\Delta_{\gamma;k} u = e^{ik\cdot x} \Delta_\gamma v$. Thus, the statement of the lemma for an arbitrary quasi-momentum k follows from the one for $k = 0$. □

3.8 Floquet–Bloch Solutions of Periodic Equations

While in the previous sections we were discussing general functions of Floquet–Bloch type, here we would like to concentrate on those that are solutions of a given periodic elliptic equation $Lu = 0$.

The next definition is a kind of tautology, but we need to formulate it anyway.

Definition 3.8.1 A solution of a periodic PDE $L(x, D)u = 0$ in $\mathbb{R}^n$, which is a Bloch (i.e., a Floquet–Bloch) function, is said to be a *Bloch* (correspondingly, a *Floquet–Bloch) solution.*

Due to ellipticity and smoothness of the coefficients of L, we do not need to worry about smoothness of solutions, since they are automatically infinitely differentiable.

We will start with providing a dual interpretation of Floquet–Bloch solutions, which happens to be a major (albeit simple) idea in many problems concerning periodic equations.

Let us consider first the case of Floquet–Bloch solutions of $Lu = 0$ with real quasi-momenta k. Any such solution u grows at most polynomially, and thus it naturally defines a continuous linear functional on $\mathcal{S}^s(\mathbb{R}^n)$ for any $s \geq 0$:

$$(u, \phi) = \int_{\mathbb{R}^n} u(x)\phi(x)\, dx, \quad \phi \in \mathcal{S}(\mathbb{R}^n). \tag{3.8.1}$$

Analogously, if we allow complex quasi-momenta, and thus exponential growth of Floquet–Bloch solutions, such solutions define by the same formula (3.8.1) continuous functionals on $H^{s,\infty}(\mathbb{R}^n)$.

The following simple, but valuable duality proposition, can be readily proven:

Proposition 3.8.2 *Let $f(x)$ be a Floquet–Bloch function in $\mathbb{R}^n$, then:*

(1) *If the quasi-momentum is real, then $f(x)$ is a solution of the mth order elliptic periodic equation $Lu = 0$ if and only if, as a functional on $\mathcal{S}^0(\mathbb{R}^n)$, it is orthogonal to the range of the dual operator $L^t : \mathcal{S}^m(\mathbb{R}^n) \to \mathcal{S}^0(\mathbb{R}^n)$ (see Section 3.7).*
(2) *If the quasi-momentum is arbitrary, then $f(x)$ is a solution of $Lu = 0$ if and only if, as a functional on $H^{0,\infty}(\mathbb{R}^n)$, it is orthogonal to the range of the dual operator $L^t : H^{m,\infty}(\mathbb{R}^n) \to H^{0,\infty}(\mathbb{R}^n)$.*

Combining this proposition with the results of Lemmas 3.6.3 and 3.6.5, one obtains the following crucial result on dual interpretation of Floquet–Bloch solutions of periodic elliptic equations (with the Floquet multipliers z used instead of quasi-momenta k):

Theorem 3.8.3

(1) *All Floquet–Bloch solutions of the equation $Lu = 0$ correspond under the Floquet transform to all "derivatives of delta"-type functionals* (3.7.5) *on the space $A(\Omega, \mathcal{E}^0)$ that are orthogonal to the range of the operator of multiplication by the operator-valued function $L^t(z)$, $z \in \Omega = (\mathbb{C} \setminus \{0\})^n$.*
(2) *All Floquet–Bloch solutions of the equation $Lu = 0$ that have real quasi-momenta correspond under the Floquet transform to all "derivatives of delta"-type functionals* (3.7.5) *on the space $C^\infty(\mathbb{T}^*, \mathcal{E}^0)$ that are orthogonal to the range of the operator of multiplication by the operator-valued function $L^t(z)$, $z \in \mathbb{T}^*$.*

In both cases, the orders of Floquet–Bloch functions are the same as the orders of the derivatives of delta-functions involved.

We will now devote some time to discussing the possible number of linearly independent Floquet solutions with a given quasi-momentum.

We will use the notation $h_{n,N}$, where

$$h_{n,N} := \binom{n+N}{N} - \binom{n+N-2}{N-2} \tag{3.8.2}$$

is the dimension of the space of all harmonic polynomials of degree $\leq N$ in n variables. We will also need the notation

$$q_{n,N} := \binom{n+N}{N} \tag{3.8.3}$$

for the dimension of the space of all polynomials of degree at most N in n variables. Notice that $q_{n-1,N}$ coincides with the dimension of the space of all homogeneous polynomials of degree N in n variables, so, in particular, $h_{n,N} = q_{n-1,N-1}+q_{n-1,N}$.

The following result provides both a crude estimate and sometimes exact dimension count. Its proof can be found in Kuchment and Pinchover (2001, 2007).

Theorem 3.8.4

(1) *Let a_{k_0} be the dimension of the kernel of the operator*

$$L(k_0) : H^m(\mathbb{T}) \to L^2(\mathbb{T}).$$

Then the dimension of the space of Floquet solutions of the equation $Lu = 0$ of order at most N with a quasi-momentum k_0 is finite and does not exceed $a_{k_0} q_{n,N}$.

(2) *Assume that zero is an eigenvalue of algebraic multiplicity 1 of the operator $L(k_0) : H^m(\mathbb{T}) \to L^2(\mathbb{T})$ and $\lambda(k)$ is the analytic branch in a neighborhood of k_0 of the simple eigenvalue of the operator $L(k)$ in $L^2(\mathbb{T})$ with the domain $H^m(\mathbb{T})$ such that $\lambda(k_0) = 0$. Let also*

$$\lambda(k) = \sum_{l=l_0}^{\infty} \lambda_l(k - k_0)$$

be the Taylor expansion of $\lambda(k)$ around the point k_0 into homogeneous polynomials, and λ_{l_0} be the first nonzero term of this expansion. Then for any $N \in \mathbb{N}$, the dimension of the space of Floquet solutions of the equation $Lu = 0$ in $\mathbb{R}^n$ of order at most N and with the quasi-momentum k_0 is equal to

$$\binom{n+N}{N} - \binom{n+N-l_0}{N-l_0}. \tag{3.8.4}$$

Remark 3.8.5 The expression in (3.8.4) coincides with the dimension of the space of all polynomials $p(x)$ in $\mathbb{R}^n$ of degree at most N, which satisfy the constant coefficient PDE

$$\lambda_{l_0}(D)p(x) = 0.$$

Moreover, one can construct an explicit isomorphism between these spaces. There is no natural isomorphism, though, and thus the isomorphism is not unique.

3.9 Comments and References

For convenience of the reader, we assemble here **some of** the main notions and results of Floquet–Bloch theory that were developed in this chapter.

- **Lattices:** If $\{\mathbf{a}_j\}$ is a basis in $\mathbb{R}^n$, integer linear combinations of its vectors $\gamma = \sum c^j \mathbf{a}_j$, $c^j \in \mathbb{Z}$ form a **(Bravais) lattice** Γ. Γ is a discrete co-compact subgroup of $\mathbb{R}^n$ (co-compactness means that Γ has a compact fundamental domain).

  If $\mathbf{b}_j$ is the biorthogonal basis to $\mathbf{a}_j$ (i.e., $\mathbf{a}_j \cdot \mathbf{b}_l = \delta_{jl}$), then linear combinations $2\pi \sum \alpha^j \mathbf{b}_j$ with integer α^j form the **reciprocal (dual)** lattice Γ^* to Γ. Elements $k \in \Gamma^*$ can be characterized by the property that

$$k \cdot \gamma \in 2\pi\mathbb{Z} \text{ for any } \gamma \in \Gamma.$$

  In our study, Γ will be the lattice of periods of the coefficients a linear partial differential operator $L(x, D) = \sum\limits_{|\alpha| \le m} a_\alpha(x) D^\alpha$.

  In the absence of Euclidean metric, the reciprocal lattice belongs to the dual, rather than the same vector space.
- **Fundamental domains:** A bounded domain $U \subset \mathbb{R}^n$ is a **fundamental domain** for the lattice Γ, if it is disjoint with its shifts $U + \gamma$ by elements γ of Γ (i.e., it contains at most one point from any Γ-orbit), and if the whole space is covered by the shifts of its closure $\widehat{U} + \gamma$.

  Fundamental domain is not unique. Two popular choices are as follows:

  (1) The fundamental parallelepiped $\sum c_j \mathbf{a}_j$, where $\{\mathbf{a}_j\}$ is a basis of Γ and all coefficients c_j take their values in $(0, 1)$.
  (2) The **Voronoi cell**, that is, the set of all points that are closer to the origin than to any other point of Γ (a popular choice in solid-state physics).

  A fundamental domain for Γ will be called **Wigner–Seitz cell** and will be denoted **W**. A fundamental domain for Γ^* will be called **Brillouin zone** and will be denoted **B**. These names are derived from solid-state physics.
- **Floquet transforms:** A Floquet transform applied to a function $f(x)$ on $\mathbb{R}^n$ produces the function

$$\mathcal{U}_\Gamma f(k,x) := \hat{f}(k,x) := \sum_{\gamma \in \Gamma} f(x+\gamma) e^{-ik\cdot\gamma}$$

  on $\mathbb{R}^n \times \mathbb{R}^n$. A more invariant, not requiring Euclidean metric, way is to define this function on $(\mathbb{R}^n)^* \times (\mathbb{R}^n)$; vectors k are called **quasi-momenta (crystal momenta)**.
- **Quasi-momenta and Floquet multipliers:** If $k \in \mathbb{C}^n$ is a quasi-momentum, then the corresponding Floquet multiplier is defined as $z = (z_1, \dots, z_n) \in \mathbb{C}^n$, where

$z_j = e^{ik\cdot a_j}$ and $\{a_j\}$ is a fixed basis of the lattice Γ. For real quasi-momenta k, the Floquet multipliers belong to the unit torus $\mathbb{T}^* \subset \mathbb{C}^n$.

- **Modified Floquet transform:** The modified Floquet transform is

$$\mathcal{V}_\Gamma f(k,x) := e^{-ik\cdot x}\mathcal{U}_\Gamma f(k,x) = e^{-ik\cdot x}\hat{f}(k,x). \tag{3.9.1}$$

The resulting function $\mathcal{V}_\Gamma f$ satisfies the following modified cyclicity conditions with respect to k and x:

$$\mathcal{V}_\Gamma f(k+\kappa,x) = e^{-i\kappa\cdot x}\mathcal{V}_\Gamma f(k,x), \kappa \in \Gamma^* \tag{3.9.2}$$

$$\mathcal{V}_\Gamma f(k,x+\gamma) = \mathcal{V}_\Gamma f(k,x), \gamma \in \Gamma. \tag{3.9.3}$$

So, the resulting functions are all x-periodic and satisfy cyclic conditions with respect to k.

In the particular case when H is the Laplacian $-\Delta$, the operators $-\Delta(k)$ are called **twisted Laplacians**. We call $H(k)$ **twisted Schrödinger operators** or (more common) **Bloch Hamiltonians**.

The advantage of dealing with the family of operators $H(k)$ rather than the single operator H is that each $H(k)$, being an elliptic operator in sections of a bundle over a **compact** manifold $\mathbb{T}$, has purely discrete spectrum, unlike the operator H, whose spectrum is continuous.

For the operator $H = -\Delta + V(x)$, the corresponding $H(k)$ acts by the same differential expression on W with the domain H^2_k, that is, with the cyclic boundary conditions. In this view, the domain of the operator analytically rotates with k, while the differential expression of H stays the same. One of its advantages is that the domain H^s_k of the operator is Γ^*-periodic with respect to k, so it can be considered as depending on the Floquet multiplier $z = (e^{ik_1}, \ldots, e^{ik_n})$ only.

Alternatively, according to (3.4.11), the operator $H(k)$ can be considered as acting, for all k, on periodic functions on $\mathbb{T}$. This is achieved by commuting with an exponential function:

$$e^{-ik\cdot x} \circ (-\Delta + V(x)) \circ e^{ik\cdot x}. \tag{3.9.4}$$

This leads to the following differential expression for $H(k)$

$$-\Delta - 2ik \cdot \nabla + k^2 + V(x) \tag{3.9.5}$$

acting on the torus $\mathbb{T}$, that is, on Γ-periodic functions. The advantage of this representation is the explicit polynomial dependence of the differential expression on the quasi-momentum and the domain being fixed. On the other hand, explicit periodicity in k is lost.

The Floquet representation can be generalized to any periodic differential operator $L(x, D)$, where $D := \frac{1}{i}\frac{\partial}{\partial x}$:

$$L(k) = L(x, D + k). \tag{3.9.6}$$

There are advantages and disadvantages of both representations: constant differential expressions with analytically varying boundary conditions versus constant domain and polynomial operator polynomial (pencil). It is thus recommended to choose them judiciously, depending on the problem.

- **Direct integral decompositions of functional spaces and of the operator H:**

$$L^2(\mathbb{R}^n) = \int_{\mathbb{T}^*}^{\oplus} L^2(W), \quad \mathcal{H}^s(\mathbb{R}^n) = \int_{\mathbb{T}^*}^{\oplus} \mathcal{H}_k^s, \quad H = \int_{\mathbb{T}^*}^{\oplus} H(k). \tag{3.9.7}$$

We will not get into any deeper discussions of the direct integral technique (see, e.g., Corwin and Greenleaf, 1990, Dixmier, 1957, Folland, 1995, Reed and Simon, 1972, Scarpellini, 1999).

In some problems of the theory of periodic PDEs, one also needs to have theorems similar to Theorem 2.3.15 for some narrower functional classes (e.g., for those where f_γ decays as an exponent of order higher than 1). The reader can find such theorems and their applications in Kuchment (1993). However, we will not need them here.

Among the most important other examples of periodic operators are, in particular, Schrödinger operators

$$\left(\frac{1}{i}\frac{\partial}{\partial x} - A(x)\right)^2 + V(x)$$

with periodic electric and magnetic potentials $V(x)$ and $A(x)$ and Maxwell operators

$$\frac{1}{\mu(x)}\nabla \times \frac{1}{\epsilon(x)}\nabla\times$$

with periodic electric permittivity $\epsilon(x)$ and magnetic permeability $\mu(x)$, as well as other popular operators of mathematical physics.

Concerning the Maxwell operators, which play the major role in the photonic crystals theory (Joannopoulos et al., 2008; Kuchment, 2001), a lot of care is needed, since both analysis and numerics of their spectra are plagued with many potholes, to large degree resulting from the operator not being elliptic by itself, only as a part of an elliptic complex (Kuchment, 2001, 2016). We again postpone the corresponding detailed considerations to Kuchment (2027).

4

Dispersion Relations: Bloch, Fermi, and Floquet Varieties

4.1 Introduction

In general, as there is no analog in higher dimensions of the 1D Hill discriminant described in Section 1.3.1, the dispersion relation (see Definition 1.3.7) takes the lead. Its definition and some properties are analogous to those in 1D, although crucial distinctions and complications arise. Thus, its study becomes much more involved, requiring serious tools from operator theory, topology, and several complex variables.

In this chapter, we introduce the basic notions and properties concerning the **dispersion relations** (also called **Bloch varieties**[1]) and **Fermi surfaces** (as well as their different incarnation, **Floquet surfaces**).

Sections 4.2 and 4.3 discuss the proof that both varieties coincide with the sets of all zeros of entire functions of finite orders. Other sections deal with the relations of these varieties with the spectrum and with traces of these varieties on the real and imaginary subspaces, as well as their other properties that are crucial for spectral theory. In particular, we discuss the irreducibility of Bloch and Fermi varieties, their stratification, singularities,[2] and some other features. Some of their additional features will be also discussed later, especially in Chapter 5, devoted to spectral theory.

The chapter ends with a condensed summary and additional comments and references in Section 4.7.

Before we start the discussion in earnest, we would like to give the reader a general idea of what these varieties are, using an analogy with the constant coefficient case. This is a "trivial" example, but it might help the reader to digest the more general notions that will become crucial when studying the spectra of periodic operators.

So, suppose we have a constant coefficient operator $H(D)$, where $H(\xi)$ is a

[1] We will use the names Bloch variety and dispersion relation interchangeably, although the correct relation between them is that the former is the graph of the latter.

[2] Van Hove singularities will be addressed in Section 5.10.

polynomial (the **symbol** of the operator) in n variables and

$$D := \frac{1}{i}\frac{\partial}{\partial x}. \tag{4.1.1}$$

It is known (e.g., Hörmander, 2005) that many properties of operator L, and in particular of the equation $Hu = 0$, are encoded into its so-called **characteristic set**, that is, the set of zeros of the symbol $H(\xi)$:

$$\Xi = \{\xi \in \mathbb{C}^n \mid H(\xi) = 0\}, \tag{4.1.2}$$

which can be also interpreted as consisting of all **wave vectors** ξ such that the plane wave $u(x) = e^{ix\cdot\xi}$ satisfies the equation $Hu = 0$. More general exponential–polynomial solutions $e^{ix\cdot\xi}p(x)$, where p is a polynomial also arise (see, e.g., the 1D case in Chapter 1). As in 1D (Theorem 1.1.1), these types of solutions are complete, in the sense that various classes of solutions of the equation $Hu = 0$ can be expanded into the integrals of these exponential (or more general exponential–polynomial) solutions over ξ in Ξ (or a subset in Ξ), with respect to a functional on Ξ acting as a "measure." See Agmon (1999, 1992); Ehrenpreis (1970); Hörmander (1990, 2005); Malgrange (1955–1956, 1963); and Palamodov (1970) for these highly non trivial results.[3]

The **Fermi surface** considered in this chapter should be viewed as an analog of the characteristic variety for periodic elliptic equations. The **Bloch variety** (= **dispersion relation**) additionally involves the spectral parameter, and thus it is an analog of the set of zeros $(\xi, \lambda) \in \mathbb{C}^{n+1}$ of the polynomial $(H(\xi) - \lambda)$.

Let now

$$H = H(x, D) = \sum_{|\alpha|\leq m} a_\alpha(x) D^\alpha \tag{4.1.3}$$

be a linear partial differential operator in $\mathbb{R}^n$ with smooth[4] Γ-periodic coefficients $a_\alpha(x)$ and

$$H(x,\xi) = \sum_{|\alpha|\leq m} a_\alpha(x)\xi^\alpha, \tag{4.1.4}$$

$$H_0(x,\xi) = \sum_{|\alpha|=m} a_\alpha(x)\xi^\alpha \tag{4.1.5}$$

be its **symbol** and **principal symbol** correspondingly.

From now on, we will assume ellipticity condition (Definition 3.6.1) that requires non-vanishing of the principal symbol $H_0(x,\xi)$ for $x \in \mathbb{R}^n$ and $\xi \neq 0 \in \mathbb{R}^n$.

From time to time, one needs to impose additional (or different) conditions on the

[3] L. Ehrenpreis called the existence of such a representation of the **fundamental principle**.

[4] Smoothness (C^∞) of coefficients will always be assumed, unless is stated otherwise. This is, as the reader will easily notice, in most cases a very significant overkill, as it will be indicated in the comments sections.

operator (its order, self-adjointness, hypoellipticity, etc.), but for now the ellipticity condition that we have just imposed will be sufficient.

In Chapter 3, we decomposed the operator H into the direct integral of a family $H(k)$ of elliptic operators on the torus $\mathbb{T}$, which polynomially depend on the quasi-momentum parameter k:

$$H(k) = H(x, D + k) = \sum_{\beta} \frac{1}{\beta!} \frac{\partial^{|\beta|} H}{\partial \xi^{\beta}}(x, D_x) k^{\beta}. \tag{4.1.6}$$

For instance, for the Schrödinger operator

$$H = -\Delta + V(x)$$

with a periodic potential V, we get

$$H(k) = (D + k)^2 + V(x) = \left(\frac{1}{i}\nabla + k\right)^2 + V(x) \tag{4.1.7}$$

$$= -\Delta - 2ik \cdot \nabla + V(x) + k^2. \tag{4.1.8}$$

As Lemma 3.6.5 asserts, the operators (4.1.6) are **Fredholm** (see Appendix B) when considered as acting between the Sobolev spaces H^{s+m} and H^s on the torus. Moreover, it will be sometimes important to know that, according to the Atiyah–Singer theorem (Atiyah and Segal, 1968; Atiyah and Singer, 1968a,b; Lawson and Michelsohn, 1989; Melrose, 1993; Palais, 1965), they have **Fredholm index**

$$\operatorname{ind} H(k) := \dim \operatorname{Ker} H(k) - \dim \operatorname{Ker} H^*(k) \tag{4.1.9}$$

equal to zero, that is,

$$\dim \operatorname{Ker} H(k) = \dim \operatorname{Ker} H^*(k) \quad (= \operatorname{codim} \operatorname{Ran} H(k)). \tag{4.1.10}$$

This is not true for general elliptic systems in vector bundles over the torus (see the references above (4.1.9)).

4.2 Dispersion Relation = Bloch Variety

We will now introduce the notion that plays an important role in solid-state theory, as well as in studying general periodic operators. In 1D case, we considered it briefly in Section 1.3.3.

We emphasize again that we will use throughout the text the names "Bloch variety" and "dispersion relation" interchangeably, albeit strictly speaking, the **Bloch variety is the graph of the dispersion relation**.

Definition 4.2.1 The **(real) Bloch variety (dispersion relation)** $B_H \subset \mathbb{R}^{n+1}$ of the operator H consists of all pairs $(k, \lambda) \in \mathbb{R}^{n+1}$ such that the operator

$$H(k) - \lambda I : H^m(\mathbb{T}) \mapsto L^2(\mathbb{T}) \tag{4.2.1}$$

is not invertible.

In other words, $(k, \lambda) \in B_H$ when λ belongs to the spectrum $\sigma(H(k))$ of the elliptic operator $H(k)$ on the torus $\mathbb{T}$.

The **(complex) Bloch variety** $B_{H,\mathbb{C}} \subset \mathbb{C}^{n+1}$ of the operator H consists of all pairs $(k, \lambda) \in \mathbb{C}^{n+1}$ such that the operator (4.2.1) is not invertible.

Since operators $H(k)$ are Fredholm and have index equal to zero, one obtains the following equivalent descriptions of B_H:

Lemma 4.2.2

(1) *The* **Bloch variety** *(real or complex) of the operator H consists of all pairs $(k, \lambda) \in \mathbb{R}^{n+1}$ (correspondingly $(k, \lambda) \in \mathbb{C}^{n+1}$), such that the equation $H(k)u = \lambda u$ has a nontrivial solution in $H^m(\mathbb{T})$.*
(2) *The Bloch variety consists of pairs (k, λ) such that the equation $H(k)u = \lambda u$ has a nontrivial Γ-periodic solution, or equivalently, the equation $Hu = \lambda u$ has a nontrivial Bloch solution with quasi-momentum k.*
(3) *The (real or complex) Bloch variety of the operator H is the graph of the multiple-valued function (= dispersion relation) that assigns to each (correspondingly real or complex) quasi-momentum $k \in \mathbb{C}^n$ the spectrum $\sigma(H(k))$ of the operator $H(k)$ on the torus:*

$$k \mapsto \sigma(H(k)). \tag{4.2.2}$$

One notices the following equality:

$$B_H = B_{H,\mathbb{C}} \cap \mathbb{R}^{n+1}.$$

Corollary 4.2.3 *The Bloch variety is Γ^*-invariant with respect to k. Thus, it is sufficient to study dispersion relation over the Brillouin zone $\mathcal{B}$ only.*

The following statement, which uses ellipticity and thus Fredholm property, is standard and will be proven in Section B.2.

Lemma 4.2.4 *For any $k \in \mathbb{C}^n$, the spectrum $\sigma(H(k)) \subset \mathbb{C}$ either is discrete, or coincides with $\mathbb{C}$.*

Remark 4.2.5

(1) There is a temptation to say that for each k, the spectrum of the operator $H(k)$ is a discrete subset of the complex plane. This is often, but not always correct.

Indeed, the pathological situation when $\sigma(H(k)) = \mathbb{C}$ can occur, as the example of the operator $e^{i\varphi}d/d\varphi$ on the unit circle shows.

If H is self-adjoint, this precludes such an anomaly and leads to the discreteness of the spectrum. Another example when discreteness happens is a second-order elliptic periodic operator in $\mathbb{R}^n$ of the form

$$H = -\sum_{i,j=1}^{n} a_{ij}(x)\partial_i\partial_j + \sum_{i=1}^{n} b_i(x)\partial_i + c(x) \tag{4.2.3}$$

with real and sufficiently smooth coefficients. More sufficient conditions can be found, for example, in Agmon (2010).

(2) For elliptic systems, there exists another possible reason for such a "strange" spectral behavior. If a system H on the torus has a nonzero Fredholm index, then clearly $(H(k) - \lambda I)$ is never invertible, and so $\sigma(H(k)) = \mathbb{C}$. This does not happen with scalar equations on the torus.

Any book on solid-state theory or photonic crystals contains many graphs of dispersion relations (e.g., Ashcroft and Mermin, 1976; Joannopoulos et al., 2008; Kittel, 1976; Ziman, 1964, 1962). In Fig. 4.1, one finds an example of (part of) the dispersion relation for Schrödinger operator with a nonzero periodic potential, drawn over the 2D Brillouin zone:

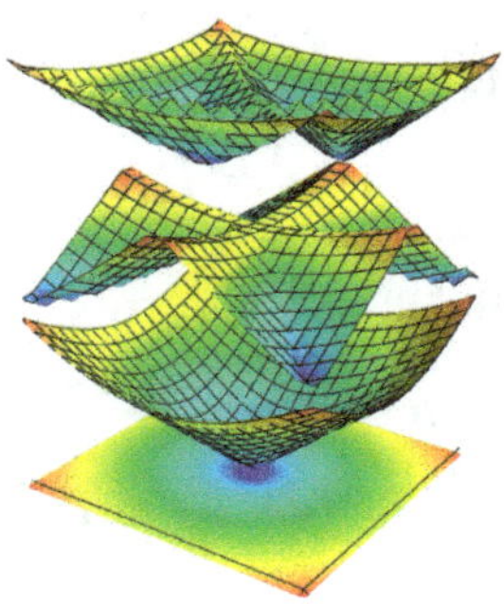

Figure 4.1 The dispersion relation (the graph of the multiple-valued function $k \to \lambda(k)$) of a non-free operator, drawn over a Brillouin zone.

Since in most practical cases the quasi-momenta are 2D or 3D, it is hard (or impossible) to visualize well the corresponding graphs. Thus, one usually draws parts of the dispersion relation graph over some special one-dimensional subsets that go along the "symmetry points" of the Brillouin zones; see Fig. 4.2. We will discuss the reasons for this approach and the risks built up in Section 4.6.2.

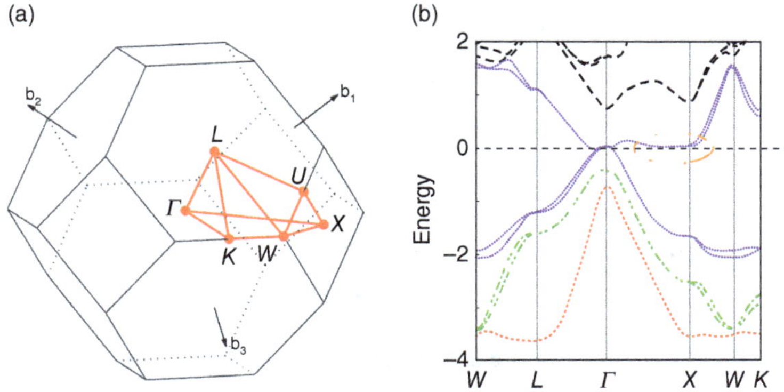

Figure 4.2 (a) A path through some symmetry points in a Brillouin zone (from Auro69, Public domain, via Wikimedia Commons). (b) The dispersion relation, drawn over a special path.

4.3 Fermi Surfaces

Let us again consider the Bloch variety as the graph of the multiple-valued function (dispersion relation) $k \mapsto \sigma(H(k))$. One might expect that the level sets of this function might be of interest. And indeed, as it will be discussed in more detail in Chapter 5 and later on, they play such an important role in physics (especially in physics of metals) that they carry the noble name the **Fermi surfaces**.[5] The notion of the Fermi surface is among the central ones, as one can conclude from any textbook on solid-state theory, for example, Ashcroft and Mermin (1976); Callaway (1964); Kittel (1976); and Ziman (1964). There are many books and databases dedicated to Fermi surfaces of various crystals, for example, Cracknell (1971); Cracknell and Wong (1973); Choy et al. (2000); Harrison and Webb (1960); Ziman (1962, 1964); and Zimbovskaya (2001). The reason is that many physical properties of crystals depend upon the geometry of their Fermi surfaces. The questions about how the analytic geometry of Fermi surfaces are intimately related to basic spectral and other properties of the relevant operator will be discussed in rest of the chapter.

To define a Fermi surface, let us consider the real (complex) dispersion relation B_H as the graph of the multiple-valued function

$$k \in \mathbb{R}^n \mapsto \sigma(H(k)), \tag{4.3.1}$$

correspondingly $k \in \mathbb{C}^n \mapsto \sigma(H(k))$.

Definition 4.3.1 The **Fermi surface** $F_{\lambda,H}$ of H at a scalar value λ is the λ-level set of the dispersion relation. That is,

[5] Also called, for obvious reasons, **isoenergy surfaces**.

Complex Fermi surface is as follows:

$$\begin{aligned} F_{\lambda,H,\mathbb{C}} &:= \{k \in \mathbb{C}^n \mid (k,\lambda) \in B_{H,\mathbb{C}}\} \\ &= \{k \in \mathbb{C}^n \mid H(k)u = \lambda u \text{ has a nonzero solution in } H_2\}. \end{aligned}$$

Real Fermi surface is as follows:

$$\begin{aligned} F_{\lambda,H} &:= \{k \in \mathbb{R}^n \mid (k,\lambda) \in B_H\} \\ &= \{k \in \mathbb{R}^n \mid H(k)u = \lambda u \text{ has a nonzero solution in } H_2\}. \end{aligned}$$

Since the Fermi surface is the level set of B_H, most properties of the Bloch variety immediately trickle down to the Fermi one. In particular, the following result follows from its analog for the Bloch variety B_H (Corollary 4.2.3):

Proposition 4.3.2 *The Fermi surface is Γ^*-periodic (invariant).*

Remark 4.3.3

(1) In physics, the name Fermi surface is used only for a specific real level of energy λ, the so-called **Fermi level**[6] λ_F (Ashcroft and Mermin, 1976; Callaway, 1964; Kittel, 1976; Ziman, 1964). It is convenient for us, though, to consider Fermi surfaces at arbitrary levels, which we hope justifies the transgression.

(2) The definition of the Fermi surface seems rather innocuous. Just to impress its possible complexity, Fig. 4.3 presents the Fermi surfaces of Niobium and Zinc. However, some part of this complexity is a kind of "illusory." Namely, let us look again at the case of a constant coefficient operator $H(D)$. The Fermi surface $F_{\lambda,H}$ in this case consists of the characteristic set $\{\xi \mid H(\xi) = \lambda\}$ **and all its Γ^* translations** (restricted then to the Brillouin zone). Thus, for instance, for the operator $H = -\Delta$ and a positive $\lambda = r^2 > 0$, it gives us the sphere

$$S_r := \{\xi \mid |\xi|^2 = r^2\} \tag{4.3.2}$$

and its shifts, then restricted to the Brillouin zone. A better way to visualize this might be to think of the sphere S_r that is cut into pieces by the dual lattice and then all these pieces are shifted back to the first Brillouin zone. Figure 4.4 might provide the reader with a better understanding of how a Fermi surface could look complex, even in simple cases.

4.3.1 Floquet Version of the Fermi Surface

What we call the **Floquet surface** is just a different coordinate incarnation of the Fermi one. Namely, since the Fermi surfaces are Γ^*-periodic, sometime one might

[6] The Fermi level is the highest energy level that an electron can occupy at the absolute zero temperature. Its existence follows from the Pauli exclusion principle.

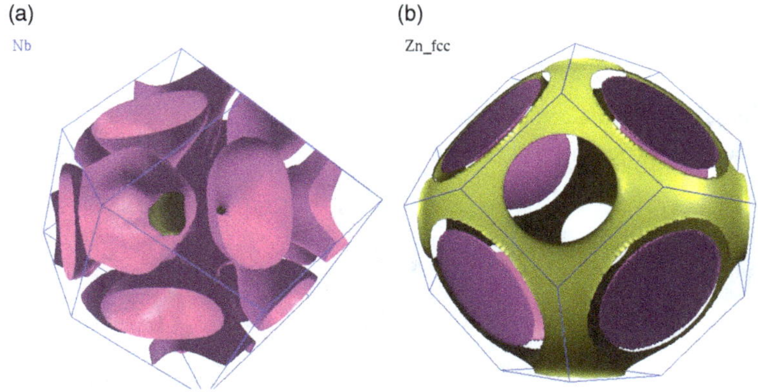

Figure 4.3 The Fermi surfaces of Niobium (a) and Zinc (b). The pictures are taken from the URL Choy et al. (2000), where one can find many more (often more complex) pictures of Fermi surfaces.

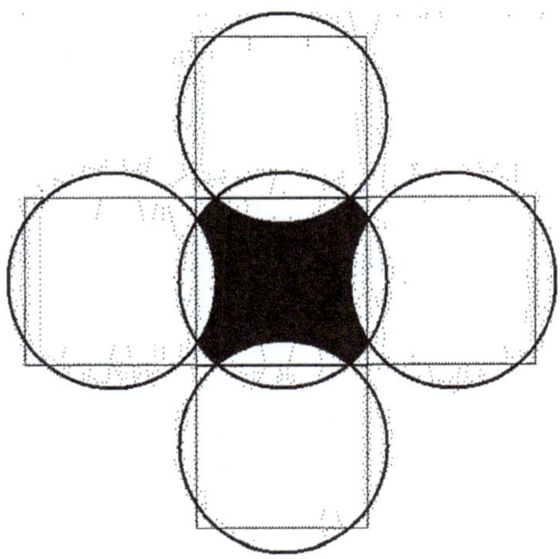

Figure 4.4 A 2D example of a Fermi surface for the Laplacian. The interior of the Fermi surface is shaded. The self-intersections correspond to the so-called Von Laue diffraction points.

want to factor out this periodicity[7] by switching from the quasi-momentum k to the Floquet multiplier variable $z = e^{ik} \in \Omega = (\mathbb{C}^*)^n$ (see Definition 3.3.2).

Definition 4.3.4 The **complex Floquet surface** $\Phi_{\lambda,H} \subset \Omega$ consists of all Floquet multipliers $z \in \Omega$ such that the equation $Hu = \lambda u$ has a nontrivial Bloch solution with the multiplier z.

[7] That is, to consider F_λ/Γ^*.

The **real Floquet surface** is defined analogously (by restricting z to the torus $\mathbb{T}^*$).

Thus, the Floquet surface at a value λ is the image of the corresponding Fermi surface under the exponential change of variables $k \to z = e^{ik}$. In other words,

$$\Phi_{\lambda,H} = F_{\lambda,H}/\Gamma^*. \tag{4.3.3}$$

Analogously,

$$\Phi_H := B_H/\Gamma^*. \tag{4.3.4}$$

The author's experience is (and the later parts of the text confirm this) that in some cases Floquet surfaces are better, while in some others the Fermi ones are preferable. Thus, a user will benefit from knowing both.

4.4 Analyticity of the Bloch and Fermi Varieties

We present now one of the most important features of the Bloch variety.

Theorem 4.4.1 *There exists an entire function $f(k,\lambda)$ on $\mathbb{C}^{n+1}$, such that*

$$B_{H,\mathbb{C}} = \{(k,\lambda) \in \mathbb{C}^{n+1} \mid f(k,\lambda) = 0\}. \tag{4.4.1}$$

This result shows that $B_{H,\mathbb{C}}$ is analytic subset of $\mathbb{C}^{n+1}$. In fact, this is what being an **analytic subset** means (see Appendix A). One can call f, the **dispersion function**.

Remark 4.4.2 For periodic PDEs of non-(hypo-)elliptic type, for example, the evolution equations of hyperbolic or Schrödinger types this property does not hold. In such cases, most of the analytic techniques discussed in this book fail and different approaches are required, as will be shown in Kuchment (2027) (see also the discussions in Kuchment, 1982, 1993, 2016).

An important stronger result holds. In order to formulate it, we remind the reader the following notion from complex analysis.

Definition 4.4.3 An entire function $f(z)$ on $\mathbb{C}^n$ is said to be of **exponential order** p if it satisfies for any $\epsilon > 0$ the estimate

$$|f(z)| \leq C_\epsilon e^{C_\epsilon |z|^{p+\epsilon}}. \tag{4.4.2}$$

Consider the complex Bloch variety

$$B_{H,\mathbb{C}} \subset \mathbb{C}^{n+1}. \tag{4.4.3}$$

Theorem 4.4.4 (Lemma 8 of §4 in Kuchment (1982); Theorem 4.4.2 of Kuchment [1993]; see also Knörrer and Trubowitz [1990])

(1) *There exists an entire function* $f(k, \lambda)$ *on* $\mathbb{C}^{n+1}$, *such that*

a.

$$|f(k, \lambda)| \leq C_p e^{(|k|+|\lambda|)^p} \tag{4.4.4}$$

for any $p > n$.

b.

$$B_{H,\mathbb{C}} = \{(k, \lambda) \in \mathbb{C}^{n+1} \mid f(k, \lambda) = 0\}. \tag{4.4.5}$$

(2) *If* $p > n + 1$, *one can make sure that* f *is* Γ^**-periodic with respect to quasi-momentum* k.

In particular, $B_{H,\mathbb{C}}$ *is a* Γ^****-periodic complex analytic sub-variety of*** $\mathbb{C}^{n+1}$ ***of co-dimension one****.*

Remark 4.4.5 A similar statement holds for more general elliptic periodic operators, with the exponent p depending on the dimension and order of the operator (Kuchment, 1993).

The proof of the theorem as stated can be found in Kuchment (1993, Theorem 4.4.2) and its sketch in Appendix F. It uses the theory of **regularized determinants of operators of Shatten–von Neumann classes**, presented, for example, in Dunford and Schwartz (1988); Gohberg and Kreĭn (1969); and Simon (2005), see also Appendix B, Section B.5, especially Theorems B.20 and B.22.

4.4.1 Analyticity of the Fermi Surface

Since the Fermi surface is just a slice of the Bloch variety, we conclude that

Theorem 4.4.6 *An analog of Theorem 4.4.4 holds for the Fermi surface* $F_{\lambda,H,\mathbb{C}}$ *for any value of* $\lambda \in \mathbb{C}$.

4.4.2 Analyticity in the Space of Parameters

An often very useful thing to do is to add to the spectral parameter λ and quasi-momenta k an finite- or infinite-dimensional space P of parameters of the periodic operator (e.g., an appropriate Banach space of periodic electric and/or magnetic potentials). Then the Fermi and Bloch varieties can be defined in this extended space. When the space of parameters is defined "properly," it is usually straightforward to prove the following:

Meta-Theorem 4.4.7 *Bloch and Fermi surfaces are analytic subsets of the enlarged space* $\mathbb{C}^{n+1} \times P$ *(respectively* $\mathbb{C}^n \times P$*).*

An example of validity of this meta-statement is the following result:

Theorem 4.4.8 *Consider the periodic Schrödinger operators $H = -\Delta + V(x)$, where $V \in L_\infty$ and the* ***extended Bloch variety***

$$\{(k, \lambda, V) \in \mathbb{C}^n \times \mathbb{C} \times L_\infty(\mathbb{T}) \mid H(k) - \lambda I \text{ is not invertible } \}. \tag{4.4.6}$$

Both Theorems 4.4.1 and 4.4.4 hold for this extended variety.

The proof of this theorem requires just a simple modification of the one for Theorem 4.4.4, taking into account the analytic dependence of the regularized determinant on the operator of the corresponding Shatten–Neumann class (Theorem B.20).

4.5 Band Functions

Let us assume that $k \in \mathbb{R}^n$ and operator H is elliptic, self-adjoint, and bounded below. These properties are inherited by the operators $H(k)$ for real k, and thus, using Lemma 4.2.4 and Remark 4.2.5, one arrives to the following statement:

Theorem 4.5.1 *Under the above conditions,*

(1) *for any $k \in \mathbb{R}^n$, the operator $H(k)$ is bounded below and has a discrete (real) spectrum*

$$\sigma(H(k)) = \{\lambda_j(k) | \lambda_1(k) \leq \lambda_2(k) \leq \cdots \leq \lambda_d(k) \leq \cdots \mapsto \infty\}; \tag{4.5.1}$$

(2) *for* ***any*** *(including complex) value of quasi-momentum k, the spectrum $\sigma(H(k))$ consists of a sequence of complex numbers $\lambda_m(k)$, $m = 1, 2, \ldots$, such that $\lim_{m\to\infty} |\lambda_m(k)| = \infty$.*

Remark 4.5.2 Notice that while for real values of the quasi-momentum one can (and does) label the eigenvalues in the nondecreasing order, **there is no natural labeling in the case of complex k.**

Recall (from (3) in Lemma 4.2.2) that a Bloch variety (= dispersion relation) is the graph of the multiple-valued function

$$k \to \lambda(k) = \sigma(H(k)) = \{\lambda_n(k)\}_{k=1}^{\infty}.$$

The branches of this multiple-valued function have their own names.

Definition 4.5.3 The function $\lambda_j(k)$ is called the ***j*th band function**.

The band functions are a natural choice of single-valued **branches** of the dispersion relation for **real** quasi-momenta. However, for complex quasi-momenta k, while the spectrum of $H = -\Delta(k) + V$ is also discrete (Agmon, 2010, pp. 180–190), **there is no natural labeling of the band functions** as λ_j.

Remark 4.5.4 It is interesting to mention that the band functions, even away from the band intersections, are not independent of each other. Namely, it is expected (and proved in dimensions 1 and 2) that the Bloch variety of a Schrödinger operator with periodic electric potential is **irreducible** (see Section 4.6.6 for this discussion). In plain words, this means that the knowledge of the values of **just one band function,** say $\lambda_3(k)$ in any small neighborhood of a quasi-momentum k_0, **determines uniquely** (by analytic continuation) **the whole dispersion relation** and thus **all band functions.**

4.6 Properties of the Band Functions

Lemma 4.6.1 *The band functions are continuous and Γ^*-periodic.*

Proof Continuity is a standard consequence of perturbation theory. Periodicity follows immediately from the definition of the Bloch variety and is established in Corollary 4.2.3. □

In fact, a much stronger statement about band functions holds:

Theorem 4.6.2 *For any m, there exists a proper (i.e., smaller than the entire) analytic subset A_m in the Brillouin zone, such that the mth band function $\lambda_m(k)$ is analytic outside A_m.*

Proof The idea of the proof is clear from the perturbation theory point of view: Band functions are analytic wherever they do not collide. However, some analysis is involved in order to prove this in the way the theorem is stated.

The graph (call it Λ_m) of $\lambda_m(k)$ over the Brillouin zone $\mathcal{B}$ is a (compact) part of the Bloch variety B_H. As such, according to Theorem 4.4.4, it is described as the set of zeros of a nontrivial entire function $f(k, \lambda)$. Let $d \in [0, \infty)$ be the order[8] of the highest derivative $\partial^d f/\partial\lambda^d$ that is identically equal to zero on Λ_m. Then, apart from the analytic set of zeros of $\partial^{d+1} f/\partial\lambda^{d+1}$, the implicit function theorem applied to the defining equation $\partial^d f/\partial\lambda^d = 0$ of Λ_m finishes the proof. □

An important feature is the following symmetry that often holds.[9]

Lemma 4.6.3 *If the operator H has real coefficients, then $\lambda(-\overline{k}) = \overline{\lambda(k)}$, where $\overline{k}$ is the vector with components complex conjugate to the ones of k. In particular, when k and λ are real, one has $\lambda(-k) = \lambda(k)$.*

[8] This is a finite number, otherwise the Bloch variety becomes the whole space, due to uniqueness of analytic continuation.

[9] This property reflects the **time reversal symmetry** for the corresponding dynamical equations. It clearly **does not hold**, for instance, **for Schrödinger equations with magnetic potentials**.

This statement is immediate, since the existence of a Bloch solution $u(x) = e^{ik\cdot x}p(x)$ of the equation $Hu = \lambda u$, in the case of real coefficients, implies that $v = \overline{u(x)} = e^{-i\overline{k}\cdot x}\overline{p(x)}$ is a solution of $Hv = \overline{\lambda}v$.

4.6.1 Spectral Bands

Although detailed considerations of the spectrum of H will be done in Chapter 5, it is convenient for us to jump ahead of ourselves a little bit and introduce some notions and results here.

We start with the **spectral bands** of the operator H.

Definition 4.6.4 The set

$$I_j := \text{range}\,(\lambda_j) = \{\lambda \in \mathbb{R} \mid \text{there exists } k \in \mathbb{R}^n \text{ such that } \lambda = \lambda_j(k)\} \quad (4.6.1)$$

is the j**th band of the spectrum** $\sigma(H)$.

Due to Γ^*-periodicity with respect to k, one can also write

$$I_j = \{\lambda \in \mathbb{R} \mid \text{there exists } k \in \mathcal{B} \text{ such that } \lambda = \lambda_j(k)\}. \quad (4.6.2)$$

There is something strange about this notion: the "spectral bands." As we defined them, they do not seem to have an obvious relation to the spectrum of the operator H. As an excuse, we notice that as we have seen in Chapter 1, in 1D this spectrum is indeed the union of the spectral bands. Chapter 5, next, will provide the complete justification for this name, showing that indeed

$$\sigma(H) = \bigcup_{j\in\mathbb{Z}^+} I_j, \quad (4.6.3)$$

and the following result holds:

Theorem 4.6.5

(1) *Each band I_j is a finite closed interval, both of which endpoints tend to infinity when $j \to \infty$.*
(2) *The bands cover the whole spectrum:*

$$\sigma(H) = \bigcup_{j\in\mathbb{Z}^+} I_j = \bigcup_{k\in\mathcal{B}} \sigma(H(k)). \quad (4.6.4)$$

(3) *Unlike the 1D case, in higher dimensions the* ***bands can*** *(and, as we will see later, "normally" do)* ***overlap****.*

4.6.2 Extrema of the Dispersion Relation

Since, according to (4.6.1)–(4.6.4), the edges of the spectrum of H occur at extrema of band functions, studying these extrema is an important task. There are several questions of interest concerning them, which we address in the following subsections.

4.6.2.1 Location

In 1D, the extrema of the dispersion relation $\lambda(k)$ occur only at $k = 0$ and $k = \pi$ in the Brillouin zone, due to the monotonicity of the band functions in the reduced Brillouin zone (Lemma 1.3.10).

Let us try to look at the points $k = 0$ and $k = \pi$ in a different light. The reciprocal lattice Γ^* is invariant with respect to the symmetry group of affine transformations generated by its shifts and "time reversal" $k \mapsto -k$. Then we find that the points $k = 0, \pi$ are exactly the fixed points of some of these symmetries inside the reduced Brillouin zone. Indeed, for instance, the affine transformation $k \mapsto -k + 2\pi$ fixes the point $k = \pi$.

We call the group of affine symmetries of a lattice that keep the origin fixed (and thus are linear) **point symmetries**. In higher dimensions, this **point symmetry group** can be large (see, e.g., Ashcroft and Mermin, 1976; Weyl, 1989). Some members of the resulting **space group** of affine symmetries generated by the lattice shifts and point symmetries might have nonzero fixed points in the Brillouin zone, similarly to the points 0 and π in 1D. If one looks at the free operator (i.e., Laplacian) case, one discovers that critical (in fact, singular) points of the overlapping paraboloids forming the Bloch variety indeed occur at symmetry points. The most practical computations for the periodic Schrödinger and Maxwell operators also showed this effect. This has led to the following discussion:

Once popular belief: *For $-\Delta + V(x)$, the extrema of the dispersion relation are attained at the points $k \in \mathcal{B}$ that are fixed by some symmetries (probably at the highest symmetry points). That is, computing the dispersion relation only at such symmetry points gives the correct spectrum as a set (although the structure of individual overlapping bands and density of spectrum would not be resolved).*

Why would one care? If this claim is correct, then in order to find the spectrum as the set, one does not need to explore the whole Brillouin zone but rather to compute spectra only along this symmetric path. This reduces the dimensionality of the heavy computation.

Frankly, there is no monotonicity (like in 1D) or any other serious reason for this claim. Alas, some numerical evidence to the contrary has been around for quite some time but was not widely known or believed. And indeed, **this claim was disproved** analytically in Exner et al. (2010) and Harrison et al. (2007) (see also simpler waveguide examples in Borisov and Pankrashkin (2013a,b)).

It is also discussed in Harrison et al. (2007) why in practice the extrema are in fact so often located at the symmetry points. The intuitive reason is that, as we have mentioned already, the extrema of the band functions for the free operator do appear at such symmetric points only. If we add a perturbation by a small periodic potential **obeying the same symmetries**, these critical points cannot move, since they are protected by the symmetry. Thus, to create a new extremum elsewhere in the Brillouin zone, one must deform the Bloch variety **very significantly** by "pushing it up or down" inside the Brillouin zone. Apparently, in most applications, this perturbation is not strong enough (although some examples have been claimed; see Adams et al., 2008).

After this discussion of the possible locations of the extrema, we now move to the probably more important questions of their structure.

4.6.2.2 Generic Structure of Spectral Edges

By a **spectral edge** we will mean a boundary point of the spectrum, i.e., the boundary point of the semi-infinite gap, or one of the two boundary points of a finite spectral gap. It is clear that such points can arise only as extrema of the dispersion relation (although not every extremum must correspond to a spectral edge).

The following complications can (and sometimes do) occur at a spectral edge value $\lambda = c$:

A. The extremal value c is attained by more than one band function.

B. The extremum c of a single band function λ_j is non-isolated (i.e., a "Mexican hat"-type picture occurs).

C. The extremum is isolated, but degenerate (i.e., the Hessian of $\lambda_j(k)$ at the extremum point is degenerate).

Although all of the previous features can occur (e.g., Filonov and Kachkovskiy, 2018, Shterenberg, 2004), the general belief is that **generically** they do not.

Conjecture 4.6.6 has been formulated in various (explicit or implicit) forms by several authors (e.g., in Avron and Simon, 1978; Colin de Verdière, 1991; Kuchment, 2016; Novikov, 1981, 1983).

Conjecture 4.6.6 *"Generically" (with respect to the potentials and other free parameters of the operator), the extrema of band functions*

(1) *are attained by a single band;*

(2) *are isolated;*

(3) *are non-degenerate, that is, the corresponding Hessians are non-degenerate.*[10]

In other words, one conjectures that generically, near such an extremum the dispersion relation looks like a single paraboloid, that is, resembles (up to a possible change of sign) the dispersion relation at the bottom of the spectrum of the free operator $-\Delta$.

Why would one make such a conjecture? Well, "philosophically," existence of a degenerate extremum of the dispersion relation is a restriction of an analytic equality type (or countably many of those). Then it is natural to expect that it generically should *not* hold for all operators. Why? Because it is hard to believe that, if such a restriction held *for all* operators, it would have been unknown.

Another intuition is that generically, the dispersion relation probably behaves like the spectrum of a "generic" family of self-adjoint matrices (which is also a nontrivial concept to understand, see, e.g., Avron and Simon, 1978; Berkolaiko and Zelenko, 2024; Von Neumann and Wigner, 1929, and references therein).

Before going any further, let us explain that the word "generic" here means a property that holds for "almost all" values of parameters. Well, and what does "almost all" mean? It depends, and there is an hierarchy of possible meanings.

Definition 4.6.7 Let P be the (Banach, or metric, or analytic, or measure, or probability) space of parameters (e.g., potentials of the operator) and $A \subset P$ be the set of values of the parameter for which the property of interest holds. When we say that the **property holds generically**, we mean one of the following **nonequivalent** conditions:

(1) The complement of A is a (proper[11]) algebraic subset of P.
(2) The complement of A is a (proper) analytic subset of P.
(3) The complement of A is the union of countably many (proper) analytic subsets of P.
(4) The complement of A is residual, that is, a countable union of closed nowhere dense sets.
(5) The complement of A has measure zero (whenever this is defined).
(6) The property holds with probability 1 (whenever this is defined).

These versions are arranged, more or less, in the order of decreasing strength of the condition imposed. In different situations, different meanings are applicable. For instance, in the discrete case, the first (algebraic) version is the most appropriate (and proven in some cases, see, e.g., Do et al., 2020). In the continuous case, the option (4) (or maybe (3)) seems to enter, while there is no hope for (1) or (2).

Although the generic validity of the conditions (1)–(3) of the Conjecture 4.6.6

[10] Non-degeneracy clearly implies the extremum being isolated.
[11] That is, smaller than the whole.

is often assumed in mathematics and physics literature (e.g., it is involved with the definition of effective masses in solid state physics, homogenization, Green's function asymptotics, Liouville-type theorems, impurity spectra in lacunas, Anderson localization), the conjecture remains largely unproven.

Here is what is known to the author in this regard.

The following classical result (Kirsch and Simon, 1987) establishes the conjecture (in a stronger form) at the bottom of the spectrum of a Schrödinger operator with periodic electric potential:

Theorem 4.6.8 (Kirsch and Simon, 1987, Theorem 2.1) *Let*

$$H = -\Delta + V(x)$$

be a periodic Schrödinger operator in $\mathbb{R}^n$. Then the bottom of the spectrum of H is attained by the nondegenerate minimum at $k = 0$ of the lowest eigenfunction λ_1 only.

More precisely, let $\lambda_0(k)$ be the lowest band and ψ be the positive periodic solution of $H\psi = \lambda_1(0)\psi$. Then

$$(a_0/a_+)^2 k^2 \leq \lambda_1(k) - \lambda_1(0) \leq k^2, \tag{4.6.5}$$

where $a_+(a_-) = \max(\min)\psi$.

An analog of this result also holds in some non-self-adjoint situations (see (6) in Theorem 4.6.41).

An improvement to Theorem 4.6.8, under additional symmetry assumption on the potential, was derived in Helffer and Hoffmann-Ostenhof (2003).

An analogous result was obtained for the periodic 2D Pauli operator (Birman and Suslina, 1999c). Regretfully, the arguments of Birman and Suslina (1999c) and Kirsch and Simon (1987) are not applicable in the case of an edge of a finite spectral gap.

Conjecture 4.6.6 was not only stated but also proven for "small" potentials in 2D in Colin de Verdière (1991):

Theorem 4.6.9 (Colin de Verdière, 1991) *For given lattice $\Gamma \subset \mathbb{R}^2$ and natural number N, there exists a residual set R in a neighborhood of zero in $C^\infty(\mathbb{T}, \mathbb{R})$, such that for any potential $V \in R$ and all $j \leq N$, the Floquet eigenvalues $\lambda_j(k)$ are simple and $\lambda_j(k)$ is a Morse function on the torus $\mathbb{T}$.*

The following partial result is known:

Theorem 4.6.10 *The claim (1) of Conjecture 4.6.6 holds for a generic Schrödinger operator with a periodic electric potential* (Klopp and Ralston, 2000) *and for divergence-type operators* (Sivaji Ganesh and Tewary, 2020).

Amazingly, the second claim of the conjecture has been proven **for all, rather than just generic** Schrödinger operators in 2D (Filonov and Kachkovskiy, 2018). It is natural to expect that this property would hold only for generic operators, rather than all, in dimensions larger than two.

In the work by Shterenberg (2004, 2005), it was shown that the presence of a magnetic potential can create degeneracies even at the bottom of the spectrum.

The conjecture also makes sense **in the discrete case**, where it does not hold true in general, but some positive results also exist, see, for example, Do et al. (2020).

Allowing, in addition to the small variations of potentials, to change the lattice of periods to a sublattice, the possibility of removing degenerate edges by such perturbations was shown in 2D (Parnovski and Shterenberg, 2017).

4.6.2.3 Dirac Cones

When two branches of the dispersion relation meet, they might form a conical junction point called a **Dirac cone** (Figure 4.5) or sometimes a "diabolic point" (Berry and Wilkinson, 1984). The former name reflects the fact that the conical structure can be locally approximated by the dispersion relation for the Dirac equation that describes massless 2D fermions (Katsnelson, 2012). In fact, wave packets concentrated near this conical point behave for some time as massless fermions (Fefferman and Weinstein, 2014).

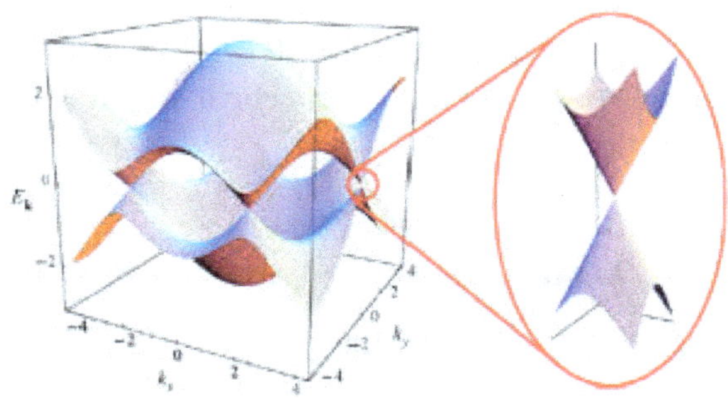

Figure 4.5 Dirac cones for the tight-binding model of graphene.

Such conical structures are often unstable under small perturbation of parameters (e.g., the potential) of the operator. Indeed, one expects that such a double-cone structure can be perturbed to become two paraboloids. However, in some cases, the conical singularity is topologically protected (see discussion and history in Berry, 1990; Thouless, 1998).

Very importantly, it has been noticed that if the structure has honeycomb lattice symmetry, which is preserved under perturbation, this protects the cone from splitting. Thus, in particular, such a cone mandatorily arises in the dispersion relation of the famous **graphene** (Figure 4.6), which explains its amazing electric properties (Katsnelson, 2012). Analogous topologically protected conical singularities arise on the boundaries of **topological insulators** (see, e.g., Bernevig, 2013). The

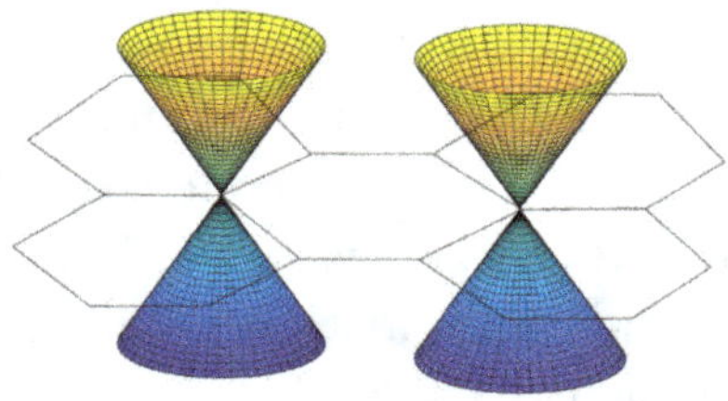

Figure 4.6 Dirac cones for the honeycomb structure.

mathematical derivation of the existence of such stable Dirac cones can be found in Fefferman and Weinstein (2012) and Berkolaiko and Comech (2018). In particular, the latter shows the existence of an irreducible representation of the symmetry group that prevents the cone from splitting during perturbation.

4.6.3 Critical Points and Singularities

Not only extrema but also other critical points and singularities of the dispersion relation as a multiple-valued function $k \mapsto \lambda(k)$ play an important role in the spectral theory, leading to the so-called **Van Hove singularities** in the spectral density (see Section 5.10).

4.6.4 On the "Size" of the Real Fermi Surface

We remind the reader of the notion of the real Fermi surface at a spectral level (energy) λ:

$$F_{\lambda,\mathbb{R},H} = \{k \in \mathcal{B} \mid \text{there exists } \Gamma\text{-periodic } u \neq 0 \text{ such that } L(k)u = \lambda u\}. \quad (4.6.6)$$

We would like to relate somehow its "size" to the location of the value λ with respect to the spectrum $\sigma(L)$. This is done to some extent in the following simple theorem:

Theorem 4.6.11

(1) $F_{\lambda,\mathbb{R},H} = \emptyset$ *if and only if* $\lambda \notin \sigma(H)$.

(2) *If* λ *belongs to the interior of a spectral band* I_n, *then* $F_{\lambda,\mathbb{R},H}$ *separates* $\mathbb{R}^n$, *and, in particular, it contains a piece which is an analytic* $(n-1)$*-dimensional submanifold of* $\mathbb{R}^n$ *or an open ball in* $\mathbb{R}^n$.

Proof The first statement is just a reformulation of Theorem 5.2.1.

Consider the second one. If λ belongs to the interior of a band I_n, this, by definition of I_n, means that the continuous function $\lambda_n(k)$ takes the value λ on $F_{\lambda,\mathbb{R},H}$, as well as some values above and below λ. This means that $F_{\lambda,\mathbb{R},H}$ separates $\mathbb{R}^n$. In addition, according to Theorem 4.3.2, $F_{\lambda,\mathbb{R},H}$ is an analytic set (i.e., the set of all zeros of an analytic function). Due to standard results on structure of analytic sets (e.g., Chirka, 1989; Grauert and Remmert, 2004; Narasimhan, 1966; and Lemma A.4), this implies the second statement of the theorem. Indeed, the real Fermi variety $F_{\lambda,\mathbb{R},H}$ can be stratified into smooth manifolds (see, e.g., Ch. V, Props. 17 and 18 of Narasimhan, 1966) and Lemma A.4. Since $F_{\lambda,\mathbb{R},H}$ separates $\mathbb{R}^n$, it must contain a stratum of dimension at least $n - 1$. These two observations imply existence of a smooth piece in $F_{\lambda,\mathbb{R},H}$ of dimension at least $(n - 1)$. □

Let us discuss this theorem a little more, as this will be useful in some other sections.

Since the real Fermi surface is the intersection of the complex one with the real subspace, we can understand the statement of Theorem 4.6.11 as when a (real) number λ is not in the spectrum, the whole Fermi surface is "hiding" in the complex domain, so the real Fermi surface is empty. When one moves λ toward the spectrum, the Fermi surface gets closer to the real space and when entering the spectrum, a real Fermi surface appears. When λ is in the interior of a spectral band, Fermi surface becomes "large," in the sense that it contains a piece of an analytic hypersurface.

There are two natural questions one might ask. First, what happens at the moment when λ is right at the edge of the spectrum? For instance, is then the real Fermi surface finite (i.e., consisting of finitely many points)? It is mentioned in Kuchment (2016), and will be discussed more in Kuchment (2027), that this question is relevant for many topics, for example, for Liouville-type theorems to hold (Kha and Kuchment, 2021; Kuchment and Pinchover, 2001, 2007). It is naturally expected that this is not necessarily correct, but is correct "generically." This issue is clearly related to the Conjecture 4.6.6.

One also wonders whether it is possible for a Fermi surface to be even larger and contain an n-dimensional rather than merely $(n-1)$-dimensional piece. The answer is given by the following rather simple, but important result, which holds in a very general setting of elliptic periodic operators:

Theorem 4.6.12 *At any energy level $\lambda \in \mathbb{R}$, the real Fermi surface either has measure zero in $\mathbb{R}^n$, or it coincides with the whole $\mathbb{R}^n$ (and thus the complex Fermi surface at this level is the whole $\mathbb{C}^n$).*

This can be rephrased as follows:

Theorem 4.6.13 ***Modified version of Theorem 4.6.12***
If the measure of the real Fermi surface $F_{\lambda_0,H}$ is positive, then the Bloch variety B_H contains the entire "flat" component

$$\mathbb{C}^n \times \{\lambda_0\} = \{(k, \lambda_0) \mid k \in \mathbb{C}^n\}.$$

Due to analyticity of the Fermi surface, the statement of Theorem 4.6.12 follows from the following uniqueness of analytic continuation statement:

Theorem 4.6.14 *Let $a(k)$ be a real analytic function on a connected open domain in $\mathbb{R}^n$. If the measure of the zero set of a is positive, then function a vanishes identically.*

Remark 4.6.15 This claim has several elementary proofs (e.g., by using implicit function theorem; Mityagin, 2020), but for some reason, it is hard to find it, even explained in most of the books on real analytic geometry (Krantz and Parks, 2002, is an exception). Due to the text size limitations, let me just indicate several other approaches, each of which can be easily completed by a graduate student, using Fubini's theorem (as suggested in Berezin and Shubin, 1991, Corollary 5.2); by induction with respect to the dimension (suggested at the beginning of Section 4.1 of Krantz and Parks, 2002); and by covering the set with a countable set of smooth submanifolds. It can also be derived from the Weierstrass Preparatory Theorem (e.g., Gunning and Rossi, 2009). It is also a direct corollary of the Lojasiewicz's result on existence of Whitney stratification of real analytic sets (Lojasiewicz, 1974, 1991), as well as of desingularization theorems in Bierstone and Milman (1988, 1991), which all seem to be quite an overkill. Another nontrivial derivation can be found in Federer (1969, Section 3.1.24, p. 240).

We will see soon the importance of the absence of such "flat" components of the Bloch variety for the spectral structure of the operator. Here is an example, a renowned result by L. Thomas (1973) (see also Section VIII.16 of Reed and Simon, 1972), although he did not formulate the statement in this form:

Theorem 4.6.16 (Thomas, 1973) *Let $V \in L_\infty(\mathbb{R}^n)$ be periodic. Then the Bloch variety of the Schrödinger operator $-\Delta + V(x)$ has no flat components $\lambda = \lambda_0$.*

Proof We assume here that $\Gamma = \mathbb{Z}^n$. The arguments can be repeated for any lattice.

Suppose that the claim of the theorem is incorrect, that is, there exists $\lambda_0 \in \mathbb{C}$ such that the equation $-\Delta(k)u + Vu = \lambda_0 u$ has a nontrivial 1-periodic solution u for any $k \in \mathbb{C}^n$. Absorbing the spectral parameter into the potential, we can assume that $\lambda_0 = 0$. Thus, we conclude that

$$\Delta(k)u = Vu \text{ has a nontrivial solution } u \text{ for any } k \in \mathbb{C}^n.$$

Let $M := \|V\|_{L_\infty}$. Then $\|Vu\|_{L^2(\mathbb{T})} \leq M\|u\|_{L^2(\mathbb{T})}$ and thus

$$\|\Delta(k)u\|_{L^2(\mathbb{T})} \leq M\|u\|_{L^2(\mathbb{T})}$$

for any k (with a nontrivial u depending on k).

Let us now pick a quasi-momentum with a large imaginary part and judiciously chosen real part. Namely, assuming that $\Gamma = \mathbb{Z}^n$ (analogous consideration works for any lattice), let

$$k = (\pi + i2a, 0, \ldots, 0), \quad a \in \mathbb{R}.$$

Expansion of the periodic function u into Fourier series easily leads to the estimate

$$\|\Delta(k)u\|_{L^2(\mathbb{T})} \geq |a|\|u\|_{L^2(\mathbb{T})}.$$

If $|a| > M$, we get a contradiction, unless $\|u\| = 0$, which we assumed does not happen. □

Remark 4.6.17

(1) Notice that the potential V in Theorem 4.6.16 does not have to be real.
(2) The assumption of boundedness of the potential is a way too strong and can be weakened (see, e.g., Section VIII.16 of Reed and Simon, 1972).
(3)

Conjecture 4.6.18 *The Bloch variety B_H for any scalar periodic* ***second-order*** *elliptic operator H with sufficiently "decent" coefficients does not have any flat components λ that are constant.*

All known counterexamples (for higher-order operators, Kuchment, 1993; or operators with insufficiently smooth coefficients in the leading terms, Filonov, 2001) are based on the failure of the weak uniqueness of continuation property.[12] Second-order elliptic equations with sufficiently "nice" coefficients, as in the previous conjecture, do have this property (see, e.g., Mizohata, 1973). This leads to a more precise conjecture:

Conjecture 4.6.19 *The Bloch variety B_H for any scalar periodic elliptic operator H that has the weak uniqueness of continuation property does not have any flat components λ that are constant.*

(4) We will see later that the absence of such flat components leads to the absolute continuity of the spectrum of the operators, which was the main goal in Thomas (1973). Due to this relation, the statements of Theorems 4.6.16 and 4.6.12 have been extended in the last three decades to a large variety of scalar and matrix periodic operators of mathematical physics (see Section 5.7.1 and references

[12] The weak uniqueness of continuation property requires that any solution vanishing on an open subset vanishes identically.

therein). A major advance was achieved in Friedlander (2002), although the proof of the full Conjecture 4.6.19 (and even of Conjecture 4.6.18 formulated later in this chapter) remains elusive.

(5) The consideration of the proof of Theorem 4.6.16 can be carried over to a variety of more general (while probably not too interesting) periodic operators $H(D)+V(x)$, where $H(D)$ is a constant coefficient operator (Kuchment, 1993, p. 28 and references therein).

(6) An interesting observation was made in Dyakin and Petrukhnovskiĭ (1988, 1982) that in the situation of Theorem 4.6.16 the Fermi surface cannot contain not only flat parts but also open pieces of some non-flat rational curves.

4.6.5 Non-algebraicity

As we have seen already, in the free case $H = -\Delta$, the Bloch variety B_H is the union of Γ^*-shifts of a single paraboloid. Thus, we get Proposition 4.6.20

Proposition 4.6.20 *In the free case,*

(1) *All irreducible components of the Bloch variety B_H are algebraic (shifted paraboloids).*

(2) *The Bloch variety B_H is irreducible*[13] *modulo Γ^*-shifts, or to put it differently, Floquet variety $\Phi_H := B_H/\Gamma^*$ is irreducible.*

The first statement of this proposition clearly holds for any constant coefficient operator $H(D)$, while the second does not.

Leaving irreducibility for Section 4.6.6, it is interesting to ask what one can expect in terms of algebraicity in the (non-free) periodic case.

In discrete periodic problems, the **Floquet varieties are algebraic** (Berkolaiko and Kuchment, 2013; Kuchment, 1991), since the corresponding operators $H(z)$ are Laurent polynomials with respect to Floquet multiplier z:

$$\sum_{j\in\mathbb{Z}^n} H_j z^j, \tag{4.6.7}$$

where the sum is finite and H_j are matrices of a finite size. Thus, multiplication by an appropriate power of z turns $H(z)$ into an operator polynomial.[14] Hence, the set of zeros of the determinant $\det(H(z) - \lambda I)$ is algebraic. This is not expected to happen in the non-free continuous case, although complete clarity in this question has not been achieved yet.

[13] See the definition and discussion of irreducibility in Section 4.6.6.

[14] Since all components of z are nonzero, such a multiplication does not influence the invertibility of the operator $H(z)$, just re-scaling correspondingly its spectrum.

In general, there appears to be no reason to expect any of the irreducible components of either the Bloch, or Floquet, variety to be algebraic
(Feldman et al., 1992).

We have already alluded to the impossibility of reducing the Floquet theory for periodic PDEs to the theory of constant coefficient PDEs, the way it was possible in the ODE case (Lyapunov's Theorem, 1.2.8). Our reasoning was that the main tool, the monodromy operator, does not work well for (non-evolutionary) PDEs. The algebraicity of Bloch variety in the constant coefficient case and non-algebraicity in the periodic case provide a more serious argument, making even the statement of the Lyapunov's theorem for elliptic PDEs questionable. Indeed, the analytic function of (k, λ) whose zeros give the Bloch variety, as Theorem 4.4.4 claims, is, as we will see in Chapter 5, a regularized infinite determinant of the polynomial operator pencil $(H(k) - \lambda I)$. This determinant is an entire function, which does not seem to have any reason whatsoever to be a polynomial. This is one of the reasons why the PDE version of Floquet theory should be developed independently of the 1D approach. Although this is not a rigorous argument, it probably persuades everyone that "generically," in the periodic case, both Bloch and Fermi varieties should not be algebraic. It is hard, though, to turn this firm belief into a rigorously proven statement ("obvious" things are often hard to prove).

The same way as for the Bloch variety, it is also expected that the Fermi surfaces of a nonconstant coefficient periodic operator are "normally" not algebraic, although we might see exceptions for some special values of λ.

We describe here the available (rather sparse) results and conjectures in this direction. The proofs are rather involved and we do not provide them, referring the reader to the original works.

Theorem 4.6.21 (Bättig et al., 1991) *Let $H = -\Delta + V(x)$ be the Schrödinger operator in $\mathbb{R}^3$ with a real $\mathbb{Z}^3$-periodic potential $V(x) \in L^2_{\text{loc}}(\mathbb{R}^3)$. If for some $\lambda \in \sigma(H)$, the real Fermi surface $F_{\lambda,\mathbb{R},H}$ contains a sphere (of non-zero diameter) or an ellipsoid, then $V(x)$ is constant.*

It is, in fact, conjectured that the mere existence of an algebraic hypersurface inside of the Fermi surface should imply that we are dealing with constant coefficients:

Conjecture 4.6.22 *If, under same conditions on the operator, for some $\lambda \in \sigma(H)$, $F_{\lambda,\mathbb{R},H}$ contains an algebraic hypersurface, then $V(x)$ is constant.*

This conjecture was proven in Bättig et al. (1991) under some additional conditions on the algebraic hypersurface. These conditions cover, in particular, the cases of spheres and ellipses, which thus implies Theorem 4.6.21.

In this relation, it is worthwhile to mention again the observation made in Dyakin and Petrukhnovskiĭ (1988, 1982) that in the situation of Theorem 4.6.16 the Fermi surface cannot contain not only flat parts but also open pieces of some non-flat rational curves.

A rather strong conjecture, proven in 2D case in Knörrer and Trubowitz (1990), is Conjecture 4.6.23:

Conjecture 4.6.23 *The following are equivalent:*

(1) *The potential is constant.*
(2) *There exists an entire function $f(k)$, whose graph $\lambda = f(k)$ belongs to the Bloch variety B_H.*

What is its importance? In 1D, it is not hard to use the known structure of the discriminant $\Delta(\lambda)$ to conclude that existence of an entire section as in Conjecture 4.6.23 is equivalent to all spectral gaps being closed. Then the famous Borg's 1D uniqueness theorem ((6) in Theorem 1.3.6) claims equivalence of this property to the potential being constant. As we will see in Chapter 5, in higher dimensions absence of spectral gaps is not an unusual feature and definitely does not imply that the potential is constant. Thus, Conjecture 4.6.23 is a way to get the "closest" **higher-dimensional analog of the Borg's theorem** for the periodic Schrödinger operator. It was proven in 2D in Knörrer and Trubowitz (1990). The discrete analog has recently been proven in arbitrary dimension in Liu (2023c).

One, however, should be aware that self-adjointness condition on the operator cannot be safely dropped in such statements. Indeed, there exist examples of **non-self-adjoint periodic operators H with algebraic components of B_H**. The following result can be found in Kuchment (1986a); see also Gasymov (1986); Veliev (1986), as well as Kuchment (1982 Theorem 11; 1993, Theorem 4.1.2).

Theorem 4.6.24 (Kuchment, 1986b; also Gasymov, 1986; and Theorem 4.1.2 in Kuchment, 1993) *Let $H = H_0(D) + H_1(x, D)$, where H_0 is an elliptic constant coefficient operator of order $2m$ in $\mathbb{R}^n$ and $H_1(x, D)$ is an operator of order at most $2m-1$ with smooth $\mathbb{Z}^n$-periodic coefficients. Suppose that there exists a convex cone $K \subsetneq \mathbb{R}^n$ with vertex at the origin, such that the Fourier series of all coefficients of H_1 contain only harmonics $e^{ik\cdot x}$ with the values of k in the interior of K. Then the Bloch variety B_H of the operator H contains the Bloch variety B_{H_0} of the constant coefficient operator H_0. Thus, B_H contains an algebraic component, albeit the coefficients of the operator are not constant.*

The idea of this theorem is very simple, it is essentially that the spectrum of an upper-triangular matrix sits on its diagonal, and the part above the diagonal is irrelevant.

Corollary 4.6.25 *If the potential $V(x)$ in the periodic Schrödinger operator $H = -\Delta + V(x)$ in $\mathbb{R}^n$ satisfies the conditions on the coefficients from Theorem 4.6.24, then the Bloch variety B_H contains the algebraic set*

$$\{(k, \lambda) \in \mathbb{C}^{n+1} \mid k^2 = \lambda^2\}. \tag{4.6.8}$$

Conjecture 4.6.26 *In both of the previous results, the Bloch varieties not only satisfy the claimed inclusion $B_{H_0} \subset B_H$, but in fact coincide: $B_{H_0} = B_H$.*

4.6.6 Irreducibility

As we will see many times throughout the book and its sequel (Kuchment, 2027; see also Kuchment, 2016), analytic properties of the Bloch and Fermi varieties play very important role in various aspects of the theory. This is not surprising, since, as we have already mentioned, these are analogs of characteristic varieties of constant coefficient operators. However, unlike the constant coefficient case, these varieties are usually more or less impossible to describe explicitly, and thus some of their analytic properties become hard to study. In Section 4.6, we will discuss one of such features, which will pop up in several spectral considerations. This is the so-called *irreducibility*. We recall the following definition (see Appendix A):

Definition 4.6.27 An analytic subset $S \in \mathbb{C}^n$ is said to be **reducible**, if it can be represented as the union of two strictly smaller analytic subsets: $S = S_1 \cup S_2$. Otherwise it is said to be **irreducible**.

Readers unfamiliar with this notion can think of two unrelated analytic functions $f_1(z)$ and $f_2(z)$ and their product $f = f_1 f_2$. Then the set S of zeros of f is the union of (usually strictly smaller) sets S_1, S_2 of zeros of f_1 and f_2. So, if f cannot be nontrivially factored (e.g., if it is an irreducible polynomial), its set of zeros is irreducible.

Another interpretation of irreducibility is the following: If one excludes all points where the set is not smooth (not an analytic manifold), irreducibility means connectedness of the remaining set of regular points (Griffiths and Harris, 1994).

We will see later on some important problems of spectral theory are intimately connected to irreducibility of the Bloch variety and Fermi surface as analytic hypersurfaces.

Let us look first at the 1D case.

Theorem 4.6.28 (Firsova, 1975; Kohn, 1959) *The Bloch variety $B(H)$ of the Hill operator $H = -\frac{d^2}{dx^2} + V(x)$ with a real $\mathbb{Z}$-periodic potential $V \in L^2_{\text{loc}}(\mathbb{R})$ is an irreducible analytic curve in $\mathbb{C}^2$.*

The proof is not that hard, one just needs to follow carefully the branching at ends (edges) of spectral bands and can be easily provided using the information from Chapter 1.

Moving now to the multidimensional situation, one expects a similar thing to happen. Namely, the following conjecture has been formulated in various publications (e.g., Avron and Simon, 1978, Knörrer and Trubowitz, 1990) and proven in the 2D case in Knörrer and Trubowitz (1990):

Conjecture 4.6.29 *For any real-valued* Γ*-periodic potential* $V(x) \in L^2_{\mathrm{loc}}(\mathbb{R}^n)$ *and* $H = -\Delta + V(x)$*, the Bloch variety* B_H *is irreducible modulo* Γ^**. In other words, the Floquet variety* $\Phi_H = B_H/\Gamma^*$ *is irreducible as an analytic hypersurface in* $\mathbb{C}^{n+1}/\Gamma^*$*.*

Remark 4.6.30

(1) The irreducibility of B_H modulo Γ^* means that any two irreducible components C_1, C_2 of B_H are related by the shift by a vector k_0 from the reciprocal lattice Γ^*:

$$C_1 + k_0 = C_2. \tag{4.6.9}$$

(2) The irreducibility property clearly implies the absence of flat components, the latter property being much weaker.

(3) The irreducibility of the Bloch variety means its **very strong rigidity**. Namely, **knowledge of any (arbitrarily small) open piece of a single branch $\lambda_j(k)$ of the dispersion relation, for example, in a small neighborhood of a point k_0, determines the whole dispersion relation**.

It is probable that the following more general conjecture also holds:

Conjecture 4.6.31 *For any elliptic* Γ*-periodic scalar second-order operator* $H(x, D)$ *in* $\mathbb{R}^n$ *with sufficiently "nice" (e.g., infinitely differentiable) coefficients, the Bloch variety* B_H *is irreducible modulo* Γ^**.*

Remark 4.6.32 The condition on the order of the operator cannot be safely removed, since, as we have discussed before, Bloch varieties of higher-order operators can contain flat pieces, and thus be reducible.

Theorem 4.6.33 (Knörrer and Trubowitz, 1990) *Conjecture 4.6.31 holds for Schrödinger operators with periodic electric potentials in 2D.*

Analogous (and usually harder to prove) conjectures have been formulated for the Fermi surfaces as well:

Conjecture 4.6.34 *For any real-valued* Γ*-periodic potential* $V(x) \in L^2_{\mathrm{loc}}(\mathbb{R}^n)$*,* $n \geq 2$*, and for (almost*[15]*) any* $\lambda \in \mathbb{C}$*, the Fermi surface* F_H *is irreducible modulo* Γ^**.*

[15] Some special values of λ can indeed produce reducible Fermi surfaces.

As in the Bloch variety case, it is probable that the following more general statement also holds.

Conjecture 4.6.35 *For any elliptic Γ-periodic second-order operator $H(x, D)$ in $\mathbb{R}^n, n \geq 2$ with sufficiently "nice" (e.g., infinitely differentiable) coefficients, and for (almost) any λ, the Fermi surface $F_{\lambda,H}$ is irreducible modulo Γ^*.*

Conjecture 4.6.34 is not that hard to prove for split potentials (Bättig et al., 1991; Kuchment and Vainberg, 2000):

Theorem 4.6.36 *Let $L = -\Delta + V_1(x_1) + V_2(x_2, x_3)$ with the potentials $V_1(x) \in L^2_{\text{loc}}(\mathbb{R})$ and $V_2(x) \in L^2_{\text{loc}}(\mathbb{R}^2)$ being real and $\mathbb{Z}$- (correspondingly $\mathbb{Z}^2$-) periodic. Then for any $\lambda \in \mathbb{C}$, the Fermi surface $F_\lambda(L)/(2\pi\mathbb{Z}^3)$ is irreducible.*

This theorem can be rather easily proven assuming the (hard) Theorem 4.6.33 from Knörrer and Trubowitz (1990). The idea of the proof is as follows: one separates variables x_1 and (x_2, x_3). The separation of variables parameter becomes a new spectral value, and the problem boils down to two Bloch varieties: one corresponding to the 1D potential V_1 and another to the 2D potential V_2. Then one uses the detailed knowledge available in 1D case and the result of Theorem 4.6.33 to obtain the irreducibility of the Fermi surface.

Having a flat component means reducibility, so this leads to a much weaker, but nontrivial and still open following conjecture:

Conjecture 4.6.37 *The Bloch variety B_H for a scalar periodic elliptic* ***second-order*** *operator H with sufficiently "decent" coefficients*[16] *does not have any flat components $\lambda =$* constant.

In self-adjoint case, this is equivalent to the absolute continuity of the spectrum.

Let us address another issue related to irreducibility. We have seen that in the free (i.e., constant coefficients) case any nonzero Γ^*-shift of an irreducible component Σ of B_H produces another component, different from Σ. In other words, irreducible components are not invariant with respect to any nonzero Γ^*-shifts. It seems that in the presence of a generic periodic potential the situation is different.

Conjecture 4.6.38 *For a generic potential, unless an irreducible component $\Sigma \subset B_H$ is algebraic, there exists a nonzero Γ^*-shift that leaves it invariant.*

If this is correct, then perturbation of the free operator by a periodic potential not only tends to deform the paraboloids but also glues them together.

[16] The condition should be sufficient for the weak uniqueness of continuation property to hold, otherwise counterexamples are known, see Filonov (2001).

4.6.7 Positive Solutions and Imaginary Bloch and Fermi Varieties below the Spectrum

Our main concentration in this chapter is on self-adjoint operators. There is, however, an important class of non-selfadjoint second order elliptic operators arising in probability theory, where somewhat different considerations are needed. We describe them briefly in Section 4.6. They will be needed when describing positive solutions of such equations (see Agmon, 1984; Lin and Pinchover, 1994), which we plan to discuss in the sequel (Kuchment, 2027).

In Section 4.6, we restrict ourselves to considering second-order elliptic periodic operators in $\mathbb{R}^n$ of the form

$$L = -\sum_{i,j=1}^{n} a_{ij}(x)\partial_i\partial_j + \sum_{i=1}^{n} b_i(x)\partial_i + c(x) \tag{4.6.10}$$

with real and sufficiently smooth coefficients. It is assumed that the inequality

$$\sum_{i,j=1}^{n} a_{ij}(x)\xi_i\xi_j \geq C\xi^2 \tag{4.6.11}$$

is satisfied for all $x, \xi \in \mathbb{R}^n$.

Such an operator, unlike most of operators h that we considered before, is not self-adjoint, unless the coefficients b_j are equal to zero. In particular, the labeling of band functions is not possible. We thus denote the operator by L, rather than H, to emphasize this difference.

The values λ of the spectral parameter that we consider here will lie "below the spectrum" of the operator L. The quotation marks are needed, since the operator is not assumed to be self-adjoint. However, in the self-adjoint case, our assumption will amount exactly to λ being below the bottom of the spectrum. We will see that for such values of the spectral parameter, it is convenient to look at the trace of the Bloch variety on the subspace of **purely imaginary** (rather than real) **quasi-momenta.**[17]

Validity of the maximum principle for the operators of the type we consider will be crucial here. The basis of what follows is the following

Theorem 4.6.39 *For every $\xi \in \mathbb{R}^n$, there exists a unique $\lambda \in \mathbb{R}$ such that the equation $Lu = \lambda u$ has a* **positive** *Bloch solution $u(x) = e^{\xi \cdot x}v(x)$ with a Γ-periodic $v(x)$. If there is another pair λ', u' with the same ξ, then $\lambda' = \lambda$ and $u'(x) = Cu(x)$.*

This theorem, whose proof can be found in Agmon (1984) and Lin and Pinchover (1994), enables us to define another object (essentially, a version of a branch of the Bloch variety).

[17] After all, Theorems 5.2.1 and 4.6.11 say that the real part of the Fermi variety would be empty, unless λ is in the spectrum.

Definition 4.6.40 The function $\Lambda(\xi) : \mathbb{R}^n \mapsto \mathbb{R}$ is defined as follows:

$$\Lambda(\xi) := \{\lambda \in \mathbb{R} | \text{ there exists a Bloch solution } u(x) = e^{\xi \cdot x}\psi(x) > 0 \text{ of } Lu = \lambda u\}. \tag{4.6.12}$$

Notice that the points $(-i\xi, \Lambda(\xi))$ of the graph of $\Lambda(\xi)$ belong to the intersection $B_L \cap \{i\mathbb{R}^n \times \mathbb{R}\}$. In fact, we will see later that this graph coincides with $B_L \cap \{i\mathbb{R}^n \times (-\infty, \Lambda_0]\}$, where

$$\Lambda_0 = \max \Lambda(\xi).$$

The next result collects the most important properties (established in Agmon, 1984, Lin and Pinchover, 1994) of this function:

Theorem 4.6.41 *The function* $\Lambda(\xi)$

(1) *is analytic,*
(2) *is bounded from above,*
(3) *is strictly concave,*
(4) *has a non-degenerate gradient, except at the maximum point,*
(5) *has non-degenerate Hessian at all points,*
(6) *satisfies when* $|\xi| \to \infty$ *the estimate*

$$c_1|\xi|^2 \leq \Lambda(\xi) \leq c_2|\xi|^2 \tag{4.6.13}$$

with positive constants c_1, c_2.

Definition 4.6.42 We denote by Λ_0 the maximal value of $\Lambda(\xi)$ and by $\xi_0 \in \mathbb{R}^n$ the (unique) point where it is attained.

It follows (see Agmon, 1984; Lin and Pinchover, 1994) that Λ_0 can also be described as follows:

$$\Lambda_0 = \sup\{\lambda \in \mathbb{R} \mid \text{there exists } u > 0 \text{ such that } (L - \lambda)u = 0 \text{ in } \mathbb{R}^n\}. \tag{4.6.14}$$

In the self-adjoint case, Λ_0 is the bottom of the spectrum of the operator L. The common name for Λ_0 is **the generalized principal eigenvalue** (also called **Agmon principal eigenvalue**) of the operator L in $\mathbb{R}^n$.

Studying the equation $Lu = 0$, one sometimes needs to know whether zero is below the spectrum or not. In other words, one wonders whether $\Lambda_0 \geq 0$ (or > 0). The next lemma combines several standard results in this regard (Lin and Pinchover, 1994). We use here the notation L^* for the operator dual to L.

Lemma 4.6.43

(1) $\Lambda_0 \geq 0$ *if and only if the equation* $Lu = 0$ *admits a positive (super-) solution. This condition is satisfied in particular when* $c(x) \geq 0$.

(2) *$\Lambda_0 \geq 0$ if and only if the equation $Lu = 0$ admits a positive Bloch solution.*

(3) *$\Lambda_0 = 0$ if and only if the equation $Lu = 0$ admits exactly one normalized positive solution in $\mathbb{R}^n$.*

(4) *If $c(x) = 0$, then $\Lambda_0 = 0$ if and only if $\int_{\mathbb{T}^n} b(x)\psi(x)\,\mathrm{d}x = 0$, where ψ is the principal eigenfunction of L^* on $\mathbb{T}^n$, and $b(x) = (b_1(x), \ldots, b_n(x))$. In particular, divergence form operators satisfy this condition.*

(5) *Let $\xi \in \mathbb{R}^n$, and assume that $u_\xi(x) = \mathrm{e}^{\xi\cdot x} p_\xi(x)$ and $u^*_{-\xi}$ are positive Bloch solutions of the equations $Lu = 0$ and $L^*u = 0$, respectively. Denote by ψ the periodic function $u_\xi u^*_{-\xi}$. Consider the function*

$$\tilde{b}_i(x) = b_i(x) - 2\sum_{j=1}^{n} a_{ij}(x)\left\{\xi_j + \left[p_\xi(x)\right]^{-1} \partial_j p_\xi(x)\right\}.$$

Then $\Lambda_0 = 0$ if and only if

$$\int_{\mathbb{T}^n} \tilde{b}_i(x)\psi(x)\,\mathrm{d}x = 0 \qquad i = 1, \ldots, n.$$

Definition 4.6.44 Suppose that $\lambda \leq \Lambda_0$, then the λ-level set of $\Lambda(\xi)$ will be denoted by $\Xi_\lambda(L)$:

$$\Xi_\lambda(L) = \{\xi \in \mathbb{R}^n \mid \Lambda(\xi) = \lambda\}.$$

In other words, $\Xi_\lambda(L)$ is (a part of) the imaginary Fermi surface for the operator L below the spectrum.

The following statement follows from the previous considerations of $\Lambda(\xi)$:

Lemma 4.6.45 *When $\lambda < \Lambda_0$, the set $\Xi_\lambda(L) \in \mathbb{R}^n$ is a compact, strictly convex, analytic hypersurface in $\mathbb{R}^n$. When $\lambda = \Lambda_0$, Ξ_λ consists of the maximum point ξ_0 only.*

The next statement specifies further the geometry of the Bloch variety below Λ_0:

Lemma 4.6.46 *Let $k = \gamma - i\xi \in \mathbb{C}^n$. If $\operatorname{Re}\lambda < \Lambda(\xi)$, then $(k, \lambda) \notin B(L)$. Moreover, if $\operatorname{Re}\lambda = \Lambda(\xi)$, then $(k, \lambda) \in B(L)$ if and only if $\gamma \in \Gamma^*$, $\xi \in \Xi_\lambda$, and* $\operatorname{Im}\lambda = 0$.

Proof Without loss of generality, we may assume that $\xi = 0$, $\Lambda(\xi) = 0$, and $L1 = 0$. Assume that $(k, \lambda) \in B(L)$, and let $u(x) = e^{i\gamma\cdot x} p(x)$ be a Floquet solution with a quasi-momentum k of the equation

$$Lu = \lambda u \qquad \text{in } \mathbb{R}^n,$$

where $\operatorname{Re}\lambda \leq 0$. Take complex conjugate:

$$L\bar{u} = \bar{\lambda}\bar{u} \qquad \text{in } \mathbb{R}^n.$$

Next, we compute

$$L(|u|^2) = \bar{u}Lu + uL\bar{u} - 2\sum_{i,j=1}^{n} a_{ij}u_{x_i}\bar{u}_{x_j} = 2z\,\mathrm{Re}\,\lambda|u|^2 - 2\sum_{i,j=1}^{n} a_{ij}u_{x_i}\bar{u}_{x_j}.$$

Notice that for each $\zeta \in \mathbb{C}^n$, we have

$$\sum_{i,j=1}^{n} a_{ij}\zeta_i\bar{\zeta}_j = \sum_{i,j=1}^{n} a_{ij}\,\mathrm{Re}\,\zeta_i\,\mathrm{Re}\,\zeta_j + \sum_{i,j=1}^{n} a_{ij}\mathrm{Im}\,\zeta_i\,\mathrm{Im}\,\zeta_j \geq 0.$$

Therefore,

$$L(|u|^2) \leq 0. \tag{4.6.15}$$

Thus, $|u|^2 = |p(x)|^2$ is a periodic nonnegative subsolution of the equation $Lu = 0$ in $\mathbb{R}^n$. By the strong maximum principle $|u(x)|^2 =$ constant. In particular, $L(|u|^2) = 0$. Since we have equality in (4.6.15), it follows that $\mathrm{Re}\,\lambda = 0$ and

$$\sum_{i,j=1}^{n} a_{ij}u_{x_i}\bar{u}_{x_j} = \sum_{i,j=1}^{n} a_{ij}\,\mathrm{Re}\,u_{x_i}\,\mathrm{Re}\,u_{x_j} + \sum_{i,j=1}^{n} a_{ij}\,\mathrm{Im}\,u_{x_i}\,\mathrm{Im}\,u_{x_j} = 0.$$

It follows that $u =$ constant and therefore, $e^{i\gamma\cdot x}$ is a periodic function. Consequently, $\gamma \in \Gamma^*$. Since $L\mathbf{1} = 0$, it follows that $\mathrm{Im}\,\lambda = 0$. □

4.7 Comments and References

- **Bloch variety**: Let $H = H(x, D)$ be a periodic elliptic operator in $\mathbb{R}^n$. The Bloch operators $H(k) = H(x, D + k)$ act on the fundamental torus $\mathbb{T}$, where k is a quasi-momentum. The *(complex) Bloch variety* of H

$$B_H = \{(k, \lambda) \in \mathbb{C}^{n+1} \mid H(k)u = \lambda u \text{ has a periodic solution } u\}. \tag{4.7.1}$$

  The *(real) Bloch variety* of L

$$B_{\mathbb{R}}(L) = B(L) \cap \mathbb{R}^{n+1}. \tag{4.7.2}$$

  Bloch variety is the graph of the multiple-valued function

$$\lambda(k) = \sigma(L(k)), \tag{4.7.3}$$

  where $\sigma(A)$ denotes the spectrum of the operator A.
- **Dispersion relation**: The multiple-valued function $\lambda(k) = \sigma(L(k))$ (for real k) is the *dispersion relation*.
- **Band functions**: Single continuous branches $\lambda_j(k)$ of the dispersion relation (usually numbered in non-decreasing order) are the *band functions*.

- **Fermi surfaces**: The *(complex) Fermi surface* $F_{\lambda,H}$ is the λ-level set of the dispersion relation:

$$F_{\lambda,H} = \{k \in \mathbb{C}^n \mid L(k)u = \lambda u \text{ has a nontrivial solution } \}. \tag{4.7.4}$$

  The *real Fermi surface* $F_{\lambda,\mathbb{R},H}$ is

$$F_{\lambda,H} \cap \mathbb{R}^n. \tag{4.7.5}$$

  The real Fermi surface $F_{\lambda,\mathbb{R},H}$ is non-empty if and only if $\lambda \in \sigma(H)$.
- **Analyticity of these varieties**: Both Bloch and Fermi varieties are sets of all zeros of entire functions of finite exponential orders. These functions can be found as appropriately regularized infinite determinants of quadratic operator pencils.
- **Section 4.3**: Understanding properties of Bloch and Fermi surfaces (e.g., irreducibility) is usually hard (see, e.g., the groundbreaking book Gieseker et al. (1993), and a shorter survey Gieseker et al. (1991), devoted solely to the case of a discrete periodic Schrödinger operator on $\mathbb{Z}^2$). In discrete case, significant progress has been achieved recently, for example, in Fisher et al. (2021); Li and Shipman (2020); Liu (2024, 2023a,b, 2022); Liu et al. (2008); and Shipman (2020).

The result of Theorem 4.4.4 not only asserts that the complex Bloch variety is an **analytic subset** (analytic hypersurface) in $\mathbb{C}^{n+1}$ (see Appendix A) but also provides an exponential bound on its defining analytic function.

Notice, for comparison, that in the constant coefficient case, the set of zeros (ξ, λ) of the polynomial $H(\xi) - \lambda$ is **algebraic**. Although there are many similarities between the properties of analytic and algebraic sets, some things get not only more complicated, but even different for the former. An example important for us is that the projection of an algebraic subset to a linear subspace is either "almost nothing" or "almost everything." More precisely, the projection either is contained in a proper algebraic subset or contains the complement of such a subset (Shafarevich, 1994, Theorem 1.25 of Section 6.3). This conclusion if an extremal importance in the case of discrete periodic equations (see, e.g., Do et al., 2020), which we do not consider here. Regretfully, such a projection result does not hold in the analytic case. However, some analogs do exist in the case of entire functions of finite exponential order (see, e.g., Alexander, 1975; Eremenko, 2006; and Julia, 1926). Thus, the estimate in Theorem 4.4.4 might (and in fact does) turn out to be crucial in some problems of spectral theory.

The reader should realize that $B_{H,\mathbb{C}}$ is not an analytic **sub-manifold** of $\mathbb{C}^{n+1}$, but rather a **sub-variety**, due to the presence of singularities (at the locations where the eigenvalue branches collide).

- **Section 4.6.2**: The observation of the mandatory appearance of Dirac cones on page 85 in honeycomb-symmetric structures was first made (way before graphene came into play) for the tight-binding model of the discrete honeycomb lattice in Wallace (1947), which provided an approximate picture of the graphene's first two dispersion bands. An infinite-band quantum graph model, which has an infinite-dimensional freedom of choosing honeycomb–symmetric potentials was considered in Kuchment and Post (2007), where detailed structure of the spectrum and dispersion relation, including in particular presence of Dirac cones, was described.[18] For 2D Schrödinger operators with honeycomb-symmetric potentials, existence and stability of Dirac cones was established, explained, and studied in Berkolaiko and Comech (2018); Fefferman et al. (2014, 2017); Fefferman and Weinstein (2012); and Grushin (2009). The Schrödinger operator with honeycomb lattice of point scatterers was considered in Lee (2016). Carbon nanotubes as folded sheets of graphene have been studied in various papers, see, for example, Do (2014); Korotyaev and Lobanov (2007); Kuchment and Post (2007); and references therein.

  One can say, however, that the story does not end here, since the presence of stable Dirac cones (including very interesting non-isotropic ones) has been observed for more general 2D (**graphyne**) structures, which lack honeycomb symmetry. See, for example, Enyanshin and Ivanovskii (2011) for physics discussion and Do (2014) and Do and Kuchment (2013) for quantum graph studies of a simplest graphyne. So far, there is no complete understanding of this phenomenon (analogous to the one provided in the honeycomb case).

  Theorem 4.6.24 requires a few pages to prove, but the meaning of this statement is very simple: Fourier expansion diagonalizes L_0, and the operator L_1 adds only an upper-triangular part, and thus does not influence the spectrum.

[18] The article Kuchment and Post (2007) also contained the detailed spectral analysis of quantum graph models of all possible types of carbon nanotubes.

5

Spectral Structure of Periodic Elliptic Operators

We have already discussed some relations between spectral theory of periodic elliptic operators and geometry of their dispersion relations and Bloch and Fermi varieties in Chapter 4. In particular, the band structure of the spectrum was reviewed. However, there is much more spectral theory of periodic operators than this, which will be analyzed in this chapter.

The reader will see that some basic questions are not completely resolved (albeit sometimes the “expected” answers are formulated).

We try to provide detailed considerations to most basic issues, while some topics (e.g., inverse problems) would have taken too much space to cover in detail. In such situations, brief surveys and references for the further reading are provided.

5.1 Essential Self-Adjointness

Although in some cases spectral theory questions can be treated in non– self-adjoint situation (see, e.g., Theorem 5.7.9), in most cases self-adjointness will be required. Thus, we will start by establishing a basic fact on essential self-adjointness of periodic elliptic operators in $\mathbb{R}^n$. As the reader can find out from the comments in Section 5.14, the result described in Theorem 5.1.1 holds in much wider generality than just periodic operators in $\mathbb{R}^n$.

Theorem 5.1.1 *Let $H(x, D)$ be an elliptic periodic operator of order m with smooth coefficients. If, defined on the space $C_0^\infty(\mathbb{R}^n)$ of smooth compactly supported functions, it is symmetric, that is,*

$$(u, Hv)_{L^2(\mathbb{R}^n)} = (Hu, v)_{L^2(\mathbb{R}^n)}, u, v \in C_0^\infty(\mathbb{R}^n),$$

then it is essentially self-adjoint, that is, has a unique self-adjoint extension. The domain of the extension is the Sobolev space $H^m(\mathbb{R}^n)$.

Proof First of all, let us denote by h the symmetric operator which is the restriction of H to $C_0^\infty(\mathbb{R}^n)$. It is clear, due to periodicity, that H acts from $H^m(\mathbb{R}^n)$ into $L^2(\mathbb{R}^n)$ as a continuous extension of h (and coincides with the action of H in the distribution sense). Since $C_0^\infty(\mathbb{R}^n)$ is dense in $H^m(\mathbb{R}^n)$, this means that the domain of the closure $\overline{h}$ of h contains $H^m(\mathbb{R}^n)$. Since h is symmetric, one has the inclusion of domains $\mathcal{D}(\overline{h}) \subset \mathcal{D}(h^*)$. So, if we show that $\mathcal{D}(h^*) \subset H^m(\mathbb{R}^n)$, this will prove the theorem. The domain of the adjoint operator h^* can be described as follows:

$$\mathcal{D}(h^*) = \{u \in L^2(\mathbb{R}^n) \mid Hu \in L^2(\mathbb{R}^n)\},$$

where H is applied to u either as the continuous extension to an operator from $L^2(\mathbb{R}^n)$ to $H^{-m}(\mathbb{R}^n)$, or equivalently, in distribution sense. Now one uses ellipticity to get local smoothness $u \in H^m_{\text{loc}}(\mathbb{R}^n)$ and periodicity to conclude that in fact $u \in H^m(\mathbb{R}^n)$ (using, for instance, Schauder estimates (Bers et al., 1979) on a periodic locally finite covering of $\mathbb{R}^n$ by bounded domains). □

5.2 The Spectrum

We will describe here the spectrum of a periodic elliptic operator of order m considered as an unbounded operator H in $L^2(\mathbb{R}^n)$ with the domain $H^m(\mathbb{R}^n)$. Notice that self-adjointness is not assumed. In Theorem 5.2.1, we assume that a Brillouin zone $\mathcal{B}$ is chosen.

Theorem 5.2.1 *The spectrum $\sigma(H)$ of the operator H coincides with the range of the dispersion relation $\lambda(k)$ over the Brillouin zone $\mathcal{B}$. That is,*

$$\sigma(H) = \bigcup_{k \in \mathcal{B}} \sigma(H(k)). \tag{5.2.1}$$

Remark 5.2.2 Recall that, due to periodicity of the dispersion relation, the union over $\mathcal{B}$ is the same as the union over the whole $\mathbb{R}^n$.

Proof Assume first that

$$\lambda \notin \bigcup_{k \in \mathcal{B}} \sigma(H(k)) \tag{5.2.2}$$

and show that then the operator H is invertible. Indeed, according to Proposition 4.3.2, the Fermi surface at the level λ, which is exactly the set of values of k where $\lambda \in \sigma(H(k))$, is analytic, and thus closed.[1] Hence, the operators $(H(k) - \lambda I)$ acting from $H^m(\mathbb{T})$ to $L^2(\mathbb{T})$ are also invertible for the values of the quasi-momentum close to the real ones. Then $(H(k) - \lambda I)^{-1}$ is a family of bounded operators in $L^2(\mathbb{T})$, analytic with respect to k in a neighborhood of $\mathbb{R}^n$. Then clearly the operator

$$\int_{\mathcal{B}}^{\oplus} (H(k) - \lambda I)^{-1} dk$$

[1] In fact, we need here only closedness rather than analyticity, and closedness is a more trivial fact.

is the bounded inverse to $(H - \lambda)$ in $L^2(\mathbb{R}^n)$. Thus, $\lambda \notin \sigma(H)$.

Suppose now that

$$\lambda \in \bigcup_{k \in \mathcal{B}} \sigma(H(k)). \tag{5.2.3}$$

We need to show that $\lambda \in \sigma(H)$. Consider

$$d := \min_{k \in \mathcal{B}} \dim \operatorname{Ker}(H(k) - \lambda I). \tag{5.2.4}$$

There are two different possibilities now: $d > 0$, or $d = 0$.

Assume first that $d > 0$ and this dimension of the kernel is attained at $k_0 \in \mathcal{B}$. Then (see Appendix B) $\operatorname{Ker}(H(k) - \lambda I)$ forms a holomorphic bundle near k_0 (i.e., there exists an analytically depending on k projector on this subspace). Hence, one can choose a continuous section $\phi(k)$ of this bundle, which is compactly supported in a neighborhood of k_0. It is clear that $\phi \in L^2(\mathcal{B}, H^m(\mathbb{T}))$ and that $H(k)\phi(k) = \lambda\phi(k)$. Then, according to Theorem 3.4.16, the inverse Floquet transform of ϕ produces an L^2-solution of $Hu = \lambda u$, and thus $\lambda \in \sigma(H)$ (in fact, $\lambda \in \sigma_{\mathrm{pp}}(H)$).

We now need to consider the case when $d = 0$. This means that the intersection of the Fermi surface with the real space is a proper (i.e., distinct from the whole space) analytic subset in $\mathbb{R}^n$. Thus, according to Theorem 4.6.12, the operator $(H(k) - \lambda I)$ is invertible for almost all (but not all) values $k \in \mathbb{R}^n$. We need to show that in this case, H is not invertible. Indeed, let k_0 be such that $\operatorname{Ker}(H(k_0) - \lambda) \neq \{0\}$. Then there exists a nonzero vector $u \in \operatorname{Ker}(H(k_0) - \lambda)$. We can continue u to a continuous $H^m(\mathbb{T})$-valued function $u(k)$ compactly supported near k_0. We can also make sure that $u(k)$ is analytic near k_0. Consider now the function $g(k) := (H(k) - \lambda I)u(k)$. By construction, it vanishes at k_0, is analytic near this point, and is not identically zero. Indeed, for almost all points k, the kernel of $(H(k) - \lambda I)$ contains only zero and thus does not contain $u(k)$. Then the function $\|g(k)\|^{-1}$ blows up at k_0. One can find a scalar function $f(k) \in L^2(\mathbb{T}^*)$ such that $\|g(k)\|^{-1} f(k)$ is not square integrable at k_0. Consider the function $\psi(k) = \|g(k)\|^{-1} f(k)u$. By construction, it is not square integrable. However, applying to it the operator

$$H - \lambda = \int_{\mathcal{B}}^{\oplus} (H(k) - \lambda I)dk, \tag{5.2.5}$$

we obtain the compactly supported function $y(k) = \|g(k)\|^{-1} f(k)g(k)$, which clearly belongs to $L^2(\mathbb{T}^*, L^2(\mathbb{T}))$. We claim that $y \notin \operatorname{Ran}(H - \lambda)$, and thus $(H - \lambda)$ is not invertible. Indeed, if y were in the range, a function $z(k) \in L^2(\mathbb{T}^*, H^m(\mathbb{T}))$ would exist, such that $(H(k) - \lambda I)z(k) = y(k)$. This, however, implies that $z(k) = \psi(k)$ almost everywhere, since the kernel of $(H(k) - \lambda I)$ is trivial for almost all values of k. Now we get the contradiction between square integrability of z, and absence of such integrability for y. □

In this theorem, no self-adjointness assumption was imposed. Let us now assume that H is symmetric, and thus accordingly to Theorem 5.1.1 is self-adjoint.

The following useful variation on the theme of the previous theorem shows that in computing the spectrum, one does not need to use all values of the quasi-momentum, but rather just those from a dense subset of the Brillouin zone:

Theorem 5.2.3 *Let H be self-adjoint and $\mathcal{B}'$ be a dense subset in the Brillouin zone $\mathcal{B}$. Then*

$$\sigma(H) = \overline{\bigcup_{k\in\mathcal{B}'} \sigma(H(k))}, \tag{5.2.6}$$

where the bar means the set closure.

Proof Assume first that $\lambda \in \overline{\bigcup_{k\in\mathcal{B}'} \sigma(H(k))}$. Then the proof of the preceding theorem applies to show that $\lambda \in \sigma(H)$.

Suppose now that $\lambda \notin \overline{\bigcup_{k\in\mathcal{B}'} \sigma(H(k))}$. Due to self-adjointness, it is sufficient to assume λ to be real. Let us denote by $\delta > 0$ the distance from λ to $\overline{\bigcup_{k\in\mathcal{B}'} \sigma(H(k))}$. Then, for any $k \in \mathcal{B}'$, the self-adjoint operator $(H(k)-\lambda I)$ on the torus is invertible in L^2. The standard estimate for resolvents of self-adjoint operators shows that the L^2 operator norm of the resolvent can be estimated as $\|(H(k)-\lambda I)^{-1}\| \le \delta^{-1}$. The operator-function $(H(k)-\lambda I)^{-1}$ is analytic outside a proper analytic set (the Fermi surface $F_{\lambda,H}$) and satisfies the uniform upper bound that we have just established on the dense set $\mathcal{B}'$. Thus, the bound holds almost everywhere on $\mathcal{B}$. Then, as before, the direct integral

$$\int_{\mathcal{B}}^{\oplus} (H(k) - \lambda I)^{-1} dk$$

provides the bounded inverse for $(H - \lambda)$. □

We will now provide a somewhat reformulated statement.

Theorem 5.2.4 *Under the same conditions on H as in Theorem 5.2.3, let $\mathcal{B}_m \subset \mathcal{B}$ be a subset of the Brillouin zone $\mathcal{B}$, which is an ϵ_m-net*[2] *in $\mathcal{B}$, where $\lim_{m\to\infty} \epsilon_m = 0$. Then, for any finite interval $\Lambda \subset \mathbb{R}$ of the spectral axis,*

$$\operatorname{dist}\left([\sigma(H)\cap\Lambda], \left[\bigcup_{k\in\mathcal{B}_m} (\sigma(H(k))\cap\Lambda)\right]\right) \to 0 \text{ when } m \to \infty. \tag{5.2.7}$$

Here dist$(\cdot,\cdot)$ *is the Hausdorff distance between sets.*

[2] An ϵ-net is a subset $U \subset \mathcal{B}$ such that any element of $\mathcal{B}$ has a representative from U in its ϵ-vicinity.

Proof According to Theorem 5.2.1, we have

$$\sigma(H) \cap \Lambda = \bigcup_{n} \bigcup_{k \in \mathcal{B}} \lambda_n(k) \cap \Lambda.$$

Now the statement follows, since only finitely many indices n are involved, the band functions λ_n are continuous, and the Brillouin zone is compact. Indeed, then one has

$$\bigcup_{n} \bigcup_{k \in \mathcal{B}} \lambda_n(k) \cap \Lambda = \lim_{m \to \infty} \bigcup_{n} \bigcup_{k \in \mathcal{B}_m} \lambda_n(k) \cap \Lambda. \tag{5.2.8}$$

□

Remark 5.2.5 The last two theorems have important uses. For instance, sometimes there are values of the quasi-momentum that are better to avoid (this is, e.g., the case with Maxwell operator, for which the value $k = 0$ is a troublesome one[3]). Namely, the domains of the operators $H(k)$ might lose their analyticity with respect to quasi-momentum. In technical terms, vector bundles turn into sheaves. Then the proof of Theorem 5.2.1 might break down. In other words, if for some reason one wants to avoid some exceptional values of k, one can do so by choosing an appropriate dense subset of quasi-momenta. In particular, this leads, due to Born and Von Karman (see, e.g., Ashcroft and Mermin, 1976), to a very useful, both theoretically and practically, way of obtaining the dispersion relation (and thus the spectrum) of a periodic operator, which we consider next.

5.2.1 Born–Karman Approximation

Our goal is to understand the spectrum $\sigma(H)$ of the operator in the whole space. This is not something that can be easily done numerically or experimentally. So, usually one chooses a "very large" domain U, imposes some boundary conditions on its boundary (say Dirichlet, Neumann, periodic, etc.), and computes the spectrum on this domain, expecting that it will be a good approximation for the spectrum in the whole space (e.g., Rozenblyum et al., 1989; Shubin, 1979). We will see now how Theorem 5.2.4 justifies such an approach.

Let us introduce some notations first. Let $N > 0$ be an integer. Consider the sub-lattice $N\Gamma \subset \Gamma$ and its fundamental domain $\mathbf{W}_N$ consisting of vectors $\sum_{j=1}^{n} \gamma^j \mathbf{a}_j$ with $\gamma^j \in [0, N]$. We can define the operator H_N in $L^2(\mathbf{W}_N)$ acting by the same differential expression (3.6.1) as H with the domain consisting of $N\Gamma$-periodic functions in $H^m(\mathbf{W}_N)$. In other words, H_N acts in the space $L^2(\mathbb{T}_N)$ with the domain $H^m(\mathbb{T}_N)$, where $\mathbb{T}_N$ is the torus $\mathbb{R}^n/(N\Gamma)$.

[3] The details will be discussed in Kuchment (2027); see also Kuchment (2001).

Theorem 5.2.6 *Under the same conditions on H as in Theorem 5.2.3, for any finite segment Λ of the spectral axis,*

$$\operatorname{dist}(\sigma(H_N) \cap \Lambda, \sigma(H) \cap \Lambda) \to 0 \text{ when } N \to \infty. \tag{5.2.9}$$

So, the spectra of the periodic problems on $\mathbf{W}_N$ approximate the spectrum of the operator in the whole space.

Proof It is sufficient to notice (using Fourier series expansions) that the functions from $H^m(\mathbb{T}_N)$ are the ones of Bloch type with the quasi-momenta $k \in \mathcal{B}_N := \frac{1}{N}\Gamma^*$. Now application of the Theorem 5.2.4 proves the statement. □

5.3 Band-Gap Structure of the Spectrum

We assume here self-adjointness and boundedness from below of the operator H, and repeat some conclusions from Chapter 4. For instance, as we discussed there, continuous, piece-wise analytic, and Γ^*-periodic **band functions** $\lambda_n(k)$ are defined. We have also defined the closed finite segments (**spectral bands**)

$$I_n := \text{ range of } \lambda_n(k) = \bigcup_{k \in \mathcal{B}} \lambda_n(k) \tag{5.3.1}$$

as the ranges of the corresponding band functions.

Definition 5.3.1 The segments I_n are called **the spectral bands** (or **allowed zones** of the spectrum) of the operator H. The open segments constituting the complement $\mathbb{R} \backslash \sigma(H)$ of the spectrum are called **gaps** (or **forbidden zones**), including one **semi-infinite gap**.

Now, the statements of Theorem 5.3.2 follows from Theorem 5.2.1 and justifies the statement of Theorem 4.6.5.

Theorem 5.3.2

(1)

$$\sigma(H) = \bigcup_n I_n, \tag{5.3.2}$$

(2) *Each band I_j is a finite closed interval, both of whose endpoints tend to infinity when $j \to \infty$.*
(3) *Every finite segment of the spectral axis intersects only finitely many bands.*
(4) *The bands cover the whole spectrum:*

$$\sigma(H) = \bigcup_{j \in \mathbb{Z}^+} I_j = \bigcup_{k \in \mathcal{B}} \sigma(H(k)). \tag{5.3.3}$$

(5) *In higher dimensions, unlike* 1*D case, the* ***spectral bands can*** *(and "normally" do)* ***overlap***.

Remark 5.3.3

(1) These conclusions are usually summarized as follows:

The spectrum $\sigma(H)$ has **band-gap structure**.

(2) The reader should notice two important things. First of all, the bands of the spectrum can, and in dimensions higher than 1, normally do overlap (e.g., see Fig. 5.1 for a simplistic representation of how the overlap can happen). The exception is the 1D case, where the 1D Schrödinger's operator theory shows that the bands cannot overlap, except touching at their ends (see Chapter 1 and references therein).

Second, while 1D spectral gaps are ubiquitous and carry a lot of information about the potential, for example, about its smoothness (see Theorem 1.3.6), existence of any gaps at higher dimensions is not a given (and definitely not a general position feature), and in fact often hard to achieve. They also carry little information about the potential. However, due to extreme importance of spectral gaps in various applications (they are responsible, for instance, for different properties of metals, insulators, and semi-conductors: see Ashcroft and Mermin, 1976), we will devote more time to discussing them.

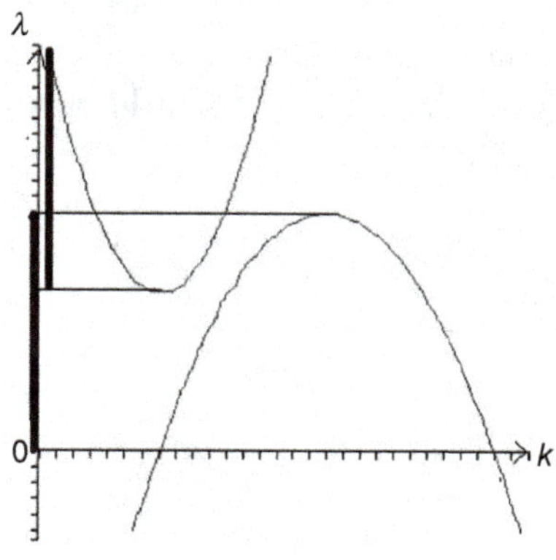

Figure 5.1 A schematic picture of band overlap.

5.4 Existence and Number of Spectral Gaps

What distinguishes self-adjoint 1D periodic second-order operators is that their spectral bands do not overlap. Thus, it stands to reason that generically all (and in particular, infinitely many) gaps should be open, which they are (see claim (4) in Theorem 1.3.6). This nice picture falls apart completely for Schrödinger operators in dimensions two and higher, where existence and number of spectral gaps do not seem to be nicely or rigidly related to the parameters of the operator (e.g., to the potential in the Schrödinger operator). This creates a lot of difficulties when gaps need to be used, created, or studied. These issues start to show up even in the simplest case of the free operator.

5.4.1 The Free Operator

So, let us start with the free operator $h = -\Delta$. As it was in 1D, it is easy to check.

Lemma 5.4.1

(1) *In any dimension, the Bloch variety B_H for the free operator $h := -\Delta$ is the union of the paraboloid $\lambda = k^2$ and its Γ^*-shifts.*
(2) *In any dimension, the spectrum of the free operator $h = -\Delta$ in $L^2(\mathbb{R}^n)$ coincides with the positive real axis, and thus has no gaps.*

Proof The proof of the first statement is done exactly as in 1D.

The second statement is well known, has no relation to periodicity, and follows by direct application of the Fourier transform.[4] □

Now one can prove the following stronger result:

Theorem 5.4.2 *In any dimension $n > 1$,* ***consecutive spectral bands of the free operator overlap****. In other words, every value $\lambda > 0$ belongs to the interior of at least one spectral band.*

Proof In order to prove this, we treat h as a $\mathbb{Z}^n$-periodic operator, so we look at its Bloch variety, that is, at solvability of the problem

$$-\Delta e^{ik\cdot x}u(x) = \lambda e^{ik\cdot x}u(x) \tag{5.4.1}$$

in $\mathbb{Z}^n$-periodic functions u (i.e., on the torus $\mathbb{T}$). Fourier series expansion

$$u(x) = \sum_{\gamma\in 2\pi\mathbb{Z}^n} u_\gamma e^{i\gamma\cdot x}$$

shows that $\lambda \in \sigma(-\Delta(k))$ if and only if there exists $\gamma \in 2\pi\mathbb{Z}^n$ such that

$$(\gamma + k)^2 = \lambda. \tag{5.4.2}$$

[4] We just pretend here that h is only a Γ-periodic operator, while it is invariant with respect to the whole group of shifts $\mathbb{R}^n$. Thus, Fourier transform applies.

The real Fermi surface, $F_{\lambda,h}$, is, according to (5.4.2), the union of spheres $S_\gamma = \{k|(\gamma + k)^2 = \lambda\}$. It is clear that on any S_γ one can find a point k that does not belong to any other sphere from that collection. This means that λ is a simple eigenvalue of the operator $h(k)$. Thus, it belongs, in a neighborhood of k, to a single branch $\lambda_j(k)$ of dispersion relation. Since branches for h are just shifted paraboloids $\lambda = k^2$, their smooth extrema can occur at $\lambda = 0$ only, and thus k is not an extremum point of λ_j. Thus, λ belongs to the interior of the corresponding band l_j. □

Remark 5.4.3 We have not developed complete understanding of the band overlaps yet. Indeed, the lengths of the overlaps could grow or decay (and indeed, in dimensions higher than three they decay to zero when $\lambda \to \infty$). When the lengths of the overlaps do decay, this creates a hope on gap opening by adding a "small" perturbation to the potential. So, the next task is to study the size of the overlap.

To quantify the overlap of the bands I_j, a continuous **overlap function** on $\mathbb{R}$ can be introduced:

Definition 5.4.4 (Skriganov, 1985a)

$$\zeta(\lambda) := \begin{cases} \max_j \max\{t \geq 0 \mid [\lambda - t, \lambda + t] \subset I_j\}, & \lambda \in \sigma(H) \\ 0, & \lambda \notin \sigma(H). \end{cases} \tag{5.4.3}$$

The following function, appearing often in number theory, turns out to be related: Let us count the number of band functions $\lambda_j(k)$ such that $\lambda_j(k) \leq \lambda$ for a fixed value of the quasi-momentum k. It is clear that the largest such j is given by

$$N(k, \lambda) := \#\{\gamma \in 2\pi\mathbb{Z}^n \mid (\gamma + k)^2 \leq \lambda\}. \tag{5.4.4}$$

One notices that $N(k, \lambda)$ is $2\pi\mathbb{Z}^n$-periodic with respect to k.

Let us introduce the following auxiliary functions:

$$\begin{aligned} N_+(\lambda) &= \max_k N(k, \lambda), \\ N_-(\lambda) &= \min_k N(k, \lambda) \end{aligned} \tag{5.4.5}$$

The function $N_+(\lambda)$ gives the maximal number j such that the jth spectral band contains a point below or equal to λ. Analogously, function $N_-(\lambda)$ produces the minimal number j such that the jth band contains a point above or equal to λ.

Thus, in particular, if we could establish for some $s > 0$, the inequality

$$N_+(\lambda - s) > N_-(\lambda + s), \tag{5.4.6}$$

this would prove that there are two overlapping zones, both containing the segment $[\lambda - s, \lambda + s]$ and thus $\zeta(\lambda) \geq 2$. In fact, the following stronger statement holds:

Theorem 5.4.5 *Let $n = 2$. Then*

(1) *for any $\lambda > 0$,*

$$N_+(\lambda) > N_-(\lambda),$$

(2) *for any $s > 0$,*

$$\lim_{\lambda\to\infty} (N_+(\lambda - s) - N_-(\lambda + s)) = \infty. \tag{5.4.7}$$

This result implies (see Parnovski, 2008; Parnovski and Sobolev, 2010; Skriganov, 1985a) the first statement of the next theorem:

Theorem 5.4.6 *For the free operator $h = -\Delta$, the following properties hold:*

(1) *In dimension $n = 2$, one has*

$$\lim_{\lambda\to\infty} \zeta(\lambda) = \infty. \tag{5.4.8}$$

(2) *When $n = 3$,*

$$\limsup_{\lambda\to\infty} \zeta(\lambda) \geq c > 0. \tag{5.4.9}$$

(3) *In dimensions $n \geq 4$, the lower bounds for the overlap functions are given by negative powers of λ.*

One can find technical details, proofs, and further references in Parnovski and Sobolev (2010) and Skriganov (1985a).

5.4.2 Periodic Operators as Perturbations of the Free Ones

If one understands the free operator that well, there is a temptation to consider the periodic potential as a perturbation term, at least at high energies $\lambda \to \infty$. This happens to be a highly valuable approach in many cases (e.g., in establishing the finiteness of the number of spectral gaps and various spectral asymptotics). For instance, if the perturbation by a potential is sufficiently "small," then it might not be strong enough to move the overlapping spectral bands sufficiently far apart. Therefore, it will be unable to remove the overlap occurring for the free operator, as described in Theorem 5.4.6, and thus to open a gap. However, looking back at Theorem 5.4.6, one can expect the difficulties of this approach to grow with the dimension, which indeed does happen. We see that, for instance, if $V \in L^\infty$, due to (5.4.8), this suffices in 2D. In 3D, due to (5.4.9), the theorem works this easy way only for potentials of sufficiently small L^∞-norms, and according to the last statement of the theorem, it does not work at all in higher dimensions. Thus, different, much more sophisticated techniques are needed. Even moderately comprehensive exposition alone could take a book (and in fact such books and surveys have been written, e.g., Karpeshina, 1997; Kuchment, 2016; Parnovski, 2008; Parnovski and Sobolev, 2010; Veliev, 2024, among others). We thus direct the reader to the original texts for the details of various techniques employed.

Let us address now the issue of the very existence of gaps. Here we immediately discover another previously mentioned difference with the 1D situation: It is easy to find in higher dimensions examples of (even non-free) periodic elliptic operators without spectral gaps (e.g., the ones in 2D or 3D with sufficiently small L_∞ norm of the potential).

Theorem 5.4.7 *Let $H = -\Delta + V(x)$ be the Schrödinger operator in $\mathbb{R}^n$ ($n = 2, 3$) with a $\mathbb{Z}^n$-periodic real potential $V \in L^\infty(\mathbb{R}^n)$. There exists a constant $C > 0$ such that if $\|V\|_{L^\infty} < C$, then the spectrum of H has no gaps.*

So, no direct analog of the Borg's uniqueness theorem (see Theorem 1.3.6) holds.

This is just an interesting corollary of Theorem 5.4.6. The proof consists of showing that in the case of homogeneous medium when $V = 0$, the spectral bands have a sizable overlap. Namely, the minimal (over the whole spectrum) length of the overlap exceeds a strictly positive constant C. Then clearly, by a perturbation argument, adding potential with the L^∞ norm smaller than C cannot move bands apart and open a gap.

Another immediate corollary in $2D$ is Theorem 5.4.8

Theorem 5.4.8 *If the periodic potential $V(x)$ of the Schrödinger operator $H = -\Delta + V(x)$ in $L^2(\mathbb{R}^2)$ is bounded, then there are only finitely many open spectral gaps.*

Indeed, if one considers H as a perturbation of the free operator $-\Delta$, the added potential cannot force gaps to open for large values of λ, where the size of the overlap exceeds the potential's L^∞-norm.

It is clear that such a simple consideration does not work in higher dimensions (and even in 3D, it can only handle small potentials).

The reader can find more discussion of and references to perturbative techniques in Section 5.14.

5.4.3 Number of Gaps: Bethe–Sommerfeld Conjecture

So, in higher dimensions, existence of spectral gaps is in general questionable. What about their possible number, if they exist at all? While in 1D the number of gaps is generically infinite (the statement (3) of Theorem 1.3.6), Theorem 5.4.8 shows that the situation is quite different in dimensions $n > 1$. Here comes the following old conjecture:

Bethe–Sommerfeld Conjecture (BSC), (Bethe and Sommerfeld, 1933),

that claims as follows:

Conjecture 5.4.9 *In dimensions higher than 1, the spectrum of the periodic Schrödinger operator $H = -\Delta + V(x)$ has at most a finite number of gaps.*

It was proven (after a long history of partial results) in Parnovski (2008), where you can also see the references to earlier results (see also Parnovski and Sobolev, 2010, and Section 5.14)).

5.4.4 Techniques of Gap Creation

Due to the discouraging (or better say "complex," in order not to judge the nature) situation with the presence of spectral gaps and at the same time their importance in applications, one can ask the natural question: Are there mechanisms that can trigger the gaps' appearance? There are a few things that have been suggested, although more study would be much welcome.

5.4.4.1 Periodic Schrödinger Operators

Let us start with the Schrödinger operator $-\Delta + V(x)$. It happens to be rather easy to "engineer" spectral gaps here. Indeed, the following procedure can be easily made precise:

(1) One adds to the free operator $-\Delta$ (or even to $\Delta + V(x)$ with a periodic $V(x)$) a deep local potential well $W(x)$, which would lead to the emergence of a bound state (eigenvalue) below the spectrum of the free operator.
(2) Repeating periodically copies of the well at large distances from each other will lead to spreading of this eigenvalue into a thin spectral band, still at a distance from the spectrum of the original spectrum, and thus create a periodic medium with a spectral gap.

Remark 5.4.10 This potential well technique fails for Maxwell and some other (e.g., divergence type) operators, where local perturbation cannot shift the bottom of the spectrum.

5.4.4.2 High Contrast

Another approach is creating "high-contrast" media, that is, those with sharp (periodic) changes in the parameters of the operator. For instance, in the case of the periodic Maxwell operator, arising in the photonic crystal theory such a parameter could be the dielectric constant. This approach succeeds in creating spectral gaps

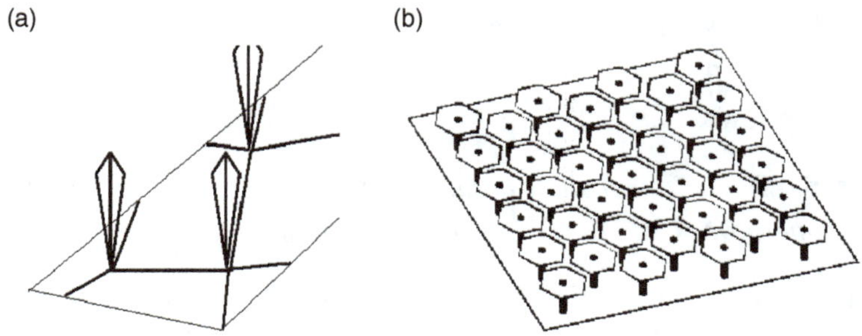

Figure 5.2 A decorated graph (a). An antenna implementing the concept (b).

for such operators (Figotin and Kuchment, 1996a,b; Filonov, 2003; Hempel and Post, 2003; Smyshlyaev and Kuchment, 2007).

Another example is creating sharp changes in the (periodic) metric tensor. This can create spectral gaps for the corresponding Laplace–Beltrami operator (Davies and Harrell, 1987; Green, 1997; Hempel and Post, 2003; Khrabustovskyi, 2012; Post, 2003, and references therein).

A two-scale homogenization approach has been developed (Zhikov, 2004) for treatment of high-contrast periodic media (see also Smyshlyaev and Kuchment, 2007).

An interesting series of papers on designing gap opening is Exner and Khrabustovskyi (2015) Khrabustovskyi (2013, 2014, 2015); and Khrabustovskyi and Khruslov (2015).

The high-contrast approach also has its drawbacks. For instance, the high-contrast approach frequently requires nonphysical values of material parameters.

5.4.4.3 Resonant Gaps

A different promising mechanism of spectral gap opening exists, which is sometimes called **resonant gaps**. The author first learned about it from Pavlov (1987). The idea is that spreading infinitely many small identical resonators throughout the medium tends to create spectral gaps around the eigenvalues of the resonator. Regretfully, to the best of the author's knowledge, this idea has not been made precise in the continuous case. It has been implemented in its simplest incarnations (as the so-called **decoration procedure**, see Fig. 5.2), in the discrete situation in Schenker and Aizenman (2000) and then in the quantum graph case explained in Berkolaiko and Kuchment (2013) and Kuchment (2005).

In practice, a more involved "spider decoration" procedures are desirable, when each vertex is replaced by a more complex structure (see Fig. 5.3 and the discussion in Berkolaiko and Kuchment (2013); Kuchment (2005); and Ong (2006)), which are at an initial stage of development.

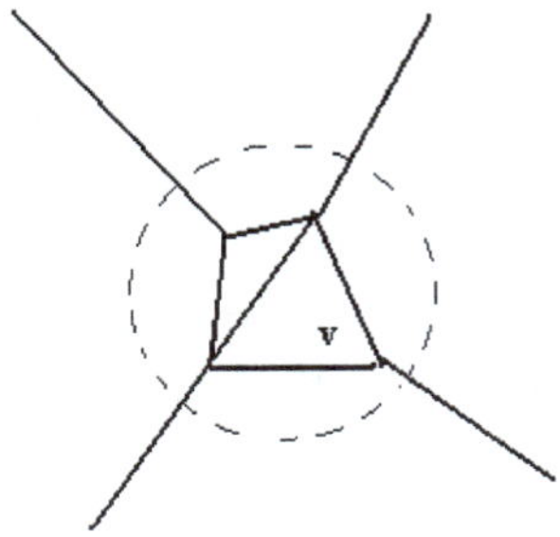

Figure 5.3 A "spider" decoration.

One can compare this with the more elaborate **zig-zag** decoration procedure used to create **expander graphs** (which boils down to controlling the size of the principal spectral gap) (Reingold et al., 2002).

Resonant gaps can also be used to create optical media that "slow down light" (e.g., Heebner et al., 2002; Kuchment, 2027, Kuchment and Kunyansky, 1999; Smyshlyaev and Kuchment, 2007; Yariv et al., 1999).

5.5 Floquet–Bloch Eigenfunctions and Spectral Decomposition

A basic issue that one has to deal with in the spectral theory of periodic operators is understanding analytic properties of Floquet–Bloch eigenfunctions as functions of the quasi-momentum and spectral parameter. As we already discussed in Theorem 5.2.1, the spectrum of a periodic elliptic operator H consists of all points λ, for which there exists a real value k of the quasi-momentum, such that the equation $Hu = \lambda u$ has a nontrivial Bloch solution $\psi_{k,\lambda}(x)$ with the quasi-momentum k. To put it differently, the equation $(H(k) - \lambda I)u = 0$ has a nontrivial periodic solution. Thus, the spectrum can be completely detected by using Bloch generalized (i.e., not square integrable) eigenfunctions. This also has led us to considering the dispersion relation, or the Bloch variety B_H, that consists of all pairs (k, λ) for which such nontrivial solution exists. We established, in particular, the important fact of analyticity of B_H (Theorem 4.4.4): It is the set of all zeros of an entire function of (k, λ) of exponential order n (dimension of the space). Bloch eigenfunctions and Bloch variety are bound to play important role in spectral theory.

One can now ask the natural question: How smoothly could the Bloch eigenfunction $\psi_{k,\lambda}(x)$ depend on (k, λ)? A closely related variation of this question is, fixing a single eigenvalue branch (which we called a **band function**) $\lambda_j(k)$, how nicely can $\psi_{k,\lambda_j(k)}(x)$ depend on k? These are the questions we are going to address and to some extent answer here.

Although in most cases, when talking about the spectral theory of periodic operators, one deals with self-adjoint setting, there are some results one can obtain without assuming self-adjointness. We will start with such a result. Let, as before $H(k)$ be the Bloch operator (3.6.11) depending on the quasi-momentum k, which corresponds to the operator H in the whole space. In other words, it is the restriction of H onto the space of functions that satisfy the Floquet–Bloch cyclicity condition (3.7.1) with the quasi-momentum k. Thus, we are interested in the kernel

$$K_{k,\lambda} := \mathrm{Ker}(H(k) - \lambda I) \tag{5.5.1}$$

depending on k.

As before, we assume for simplicity the coefficients of the operator H to be sufficiently smooth.

Theorem 5.5.1 *Let $B_H = \bigcup_l B_l$ be the (locally finite) decomposition of the Bloch variety into irreducible components $B_l \subset \mathbb{C}^{n+1}$ (see Appendix A). Then, for each l there exists a proper analytic subset A_l of B_l of dimension $(n-1)$, and a finite set of functions*

$$\psi_{j,l,(k,\lambda)}(x), \quad j = 1, \dots, j_m \tag{5.5.2}$$

such that

(1) *The functions $\psi_{j,l,(k,\lambda)}(x)$ are analytic on $B_l \setminus A_l$ as functions of (k, λ) with values in the Sobolev space $H^s(\mathbb{T})$.*
(2) *For each $(k, \lambda) \in B_l \setminus A_l$, $\psi_{j,l,(k,\lambda)}(x) \in K_{k,\lambda}$.*
(3) *For each $(k, \lambda) \in B_l \setminus A_l$, the functions $\psi_{j,l,(k,\lambda)}(x)$, $j = 1, \dots, j_m$ form a basis in $K_{k,\lambda}$.*

In particular, the fibration

$$\coprod_{(k,\lambda)\in B_l\setminus A_l} K_{k,\lambda} \mapsto (B_l \setminus A_l) \tag{5.5.3}$$

is a trivializable holomorphic vector sub-bundle of the trivial bundle over $(B_l \setminus A_l)$ with the fiber $H^s(\mathbb{T})$.

The proof of this theorem, albeit not too hard, is technical and involves tools from the several complex variables theory. It is proved in Appendix F.

Thus, the theorem claims a possibility of choosing (for a given irreducible component) a finite basis of Bloch eigenfunctions analytically depending on the quasi-momentum and eigenvalue, albeit the basis property fails on a "thin" subset A. Thinness means here that A is analytic and of smaller dimension than the whole component of the dispersion relation.

However, since this is done for each irreducible component, and thus if the number of components happens to be infinite, one might end up with an infinite set of Bloch eigenfunctions needed. On the other hand, it is known for self-adjoint Schrödinger operators in $2D$ (Theorem 4.6.33) and conjectured for Schrödinger and other second-order operators in higher dimensions (Conjectures 4.6.29 and 4.6.31) that B_H is irreducible. In such cases, Theorem 5.5.1 guarantees **existence of a single finite analytic family of Bloch eigenfunctions** generating the kernel almost everywhere on the dispersion curve.

The Gram–Schmidt orthonormalization procedure now can be applied wherever the basis property holds. This, however, will lead to discontinuities along the sets where this property fails, as well as to smooth, rather than analytic behavior in "good" parts. Thus, one gets the following corollary.

Corollary 5.5.2 *For any irreducible component B_l, one can choose Bloch eigenfunctions $\phi_{j,l,(k,\lambda)}$ such that they form an orthonormal basis in $K_{k,\lambda}$ and are smooth functions of $(k, \lambda) \in B_l$ outside of an analytic subset $A \subset B_l$.*

Let us now assume that the operator H is self-adjoint and restrict our attention to the real quasi-momenta only. We will also look at a single band function $\lambda_l(k)$ for k in the Brillouin zone. We then immediately obtain the following corollary.

Corollary 5.5.3 *For any band function $\lambda_l(k)$, there exists a proper analytic subset A_l in the Brillouin zone $\mathcal{B}$ and a finite family of Bloch eigenfunctions $\phi_{j,k}$ analytic with respect to $k \in \mathcal{B} \setminus A_l$ such that they form a basis in $K_{k,\lambda_l(k)}$ for $k \in \mathcal{B} \setminus A_l$. One can assume the basis to be orthonormal, at the expense of having only piecewise-smooth dependence on k, with possible discontinuities along A_l. In particular, one achieves measurable dependence of the orthonormal basis on the quasi-momentum.*

Remark 5.5.4 In the previous statements, a finite number of Bloch eigenfunctions is mentioned. However, if one needs more than one such function, this means that along the whole irreducible component the eigenvalue is multiple. The author is unaware of such examples of scalar periodic PDEs and formulates (with some hesitation) the following conjecture.

Conjecture 5.5.5 *Such multiplicity situation can never arise for the second-order scalar operators,*[5] *and thus only one analytic Bloch eigenfunction is needed.*

This is the case, for instance, for periodic Schrödinger operators in $2D$, if there is at least one eigenvalue $\lambda_l(k_0)$ that is simple, that is, if for some (k, λ) one has $\dim \operatorname{Ker}(H(k) - \lambda I) = 1$. Indeed, Theorem 4.6.33 guarantees irreducibility of the Bloch variety in this case.

[5] Or at least for Schrödinger operators with periodic electric potential.

5.6 Orthonormalization

The orthonormal property is usually required for the real values of quasi-momenta only. This allows one to do orthonormalization for real k in a manner analytic in k. Indeed, let U be a neighborhood in $\mathbb{C}^n$ of the space of real quasi-momenta $\mathbb{R}^n$. We will assume that U is symmetric with respect to the complex conjugation $k = (k_1, \ldots, k_n) \mapsto \overline{k} = (\overline{k_1}, \ldots, \overline{k_n})$. Let $\phi(k)$ be an analytic function in U with values in a complex Hilbert space H ($H = L^2(W)$ will be our usual case) with the Hermitian scalar product $(\cdot, \cdot)$ (which is thus anti-linear with respect to the second factor). Then the function $(\phi(k), \phi(\overline{k}))$ is analytic in U and coincides with $\|\phi(k)\|^2$ for real k. If now $\|\phi(k)\|$ does not vanish for real k, one can normalize $\phi(k)$ as follows:

$$\psi(k) := \frac{\phi(k)}{\sqrt{(\phi(k), \phi(\overline{k}))}}. \tag{5.6.1}$$

Then $\psi(k)$ is analytic in a neighborhood of the real space and has unit norm for real values of k. Notice that the expression in (5.6.1) can lose its analyticity at some point, due to the denominator, and thus the neighborhood U might need to shrink somewhat.

Applying the Gram–Schmidt procedure in this manner, one gets the following result:

Theorem 5.6.1 *Let U be a neighborhood on $\mathbb{R}^n \subset \mathbb{C}^n$ and and let the functions $\phi_1(k), \ldots, \phi_m(k)$ with values in a complex Hilbert space H be analytic in U and linearly independent on $U \setminus A$, where $A \subset U$ is an analytic subset. Then there exist a neighborhood $\mathbb{R}^n \subset V \subset U$, an analytic set $B \supset A$, and a family of functions $\psi_1(k), \ldots, \psi_m(k)$ such that they are analytic in $V \setminus B$, form a basis in the span of $\phi_1(k), \ldots, \phi_m(k)$ for $k \in V \setminus B$, and are orthonormal for $k \in \mathbb{R}^n \setminus B$.*

The immediate consequence of this result is the following generalization of Corollary 5.5.3:

Corollary 5.6.2 *For any band function $\lambda_l(k)$, there exists a proper analytic subset A in a complex neighborhood U of the space $\mathbb{R}^n$ of real quasi-momenta and a finite family of Bloch eigenfunctions $\phi_{j,k}$ analytic in $k \in U \setminus A$ such that they form a basis in $K_{k,\lambda_l(k)}$ for $k \in U \setminus A$, the basis being orthonormal for real values of quasi-momenta, that is, for $k \in \mathbb{R}^n \setminus A$.*

We are ready now to prove the spectral decomposition theorem from Gel′fand (1950). Let, as in the previous section, H be a self-adjoint periodic elliptic operator. Let us choose for each band function $\lambda_m(k)$ a piecewise smooth with respect to k orthonormal basis $\phi_{j,m}(k, x)$ of Bloch eigenfunctions, where $j = 1, \ldots, l_m$.

Theorem 5.6.3 *Let $f(x) \in L^2(\mathbb{R}^n)$ and*

$$\alpha_{j,m}(k) = \int_{\mathbb{R}^n} f(x)\overline{\phi_{j,m}(k,x)}dx, \tag{5.6.2}$$

where the integral is understood in the sense of L^2-convergence when $N \to \infty$ of the truncated integrals over cubes $[-N,N]^n$. Then,

(1) $\int_{\mathbb{R}^n} |f(x)|^2 dx = \sum\limits_{j,m} \int_{\mathcal{B}} |\alpha_{j,m}(k)|^2\, dk.$

(2) $f(x) = \sum\limits_{j,m} \int_{\mathcal{B}} \alpha_{j,m}(k)\phi_{j,m}(k,x)\, dk.$

(3) *If f belongs to the domain of H, then*

$$Hf(x) = \sum_{j,m} \int_{\mathcal{B}} \lambda_m(k)\alpha_{j,m}(k)\phi_{j,m}(k,x)\, dk.$$

Here $\mathcal{B}$ is the Brillouin zone.

In other words, operator H is unitarily equivalent to the orthogonal direct sum of operators of multiplication by piece-wise analytic continuous scalar functions $\lambda_j(k)$ in $L^2(\mathbb{T}^*)$.

Proof Let $\mathbf{W}$ be the Wigner–Seitz cell. We can split the integration in (5.6.2) into integration over the shifted cells $\mathbf{W} + \gamma$, where $\gamma \in \Gamma = \mathbb{Z}^n$:

$$\begin{aligned}\alpha_{j,m}(k) &= \sum_{\gamma\in\Gamma} \int_{\mathbf{W}+\gamma} f(x)\overline{\phi_{j,m}(k,x)}\, dx \\ &= \sum_{\gamma\in\Gamma} \int_{\mathbf{W}} f(x)e^{-ik\cdot\gamma}\overline{\phi_{j,m}(k,x)}\, dx \\ &= \int_{\mathbf{W}} (\mathcal{U}_\Gamma f)(k,x)\overline{\phi_{j,m}(k,x)}\, dx.\end{aligned}$$

Since the Bloch eigenfunctions form an orthonormal basis for almost all values of quasi-momenta, one gets

$$\|(\mathcal{U}_\Gamma f)(k,\cdot)\|^2_{L^2(\mathbb{T})} = \sum_m |\alpha_{j,m}(k)|^2.$$

Now integration over the Brillouin zone and using Theorem 3.4.1 finish the proof of the first statement of the theorem.

The right-hand side expression in the second statement of the theorem is equal to $\int_{\mathcal{B}} (\mathcal{U}_\Gamma f)(k,x)$ and thus to

$$\sum_{\gamma\in\Gamma} f(x+\gamma) \int_{\mathcal{B}} e^{-i\gamma\cdot k} dk.$$

All terms of this sum vanish, except the one with $\gamma = 0$, which proves the second statement.

We leave the simple proof of the third statement to the readers. □

5.7 Classification of the Spectra

Let H, as before, be a self-adjoint periodic elliptic operator in $\mathbb{R}^n$ with (sufficiently) smooth coefficients. We would like now to look at the presence of various parts in the standard decomposition of its spectrum into the pure point, absolutely continuous, and singular continuous parts (see Section B.2 and the references therein):

$$\sigma(H) = \sigma_{\text{pp}}(H) \cup \sigma_{\text{ac}}(H) \cup \sigma_{\text{sc}}(H). \tag{5.7.1}$$

According to Theorem 5.6.3, it is sufficient to understand the structure of the spectrum of the operator A acting in $L^2(\mathbb{T}^*)$ as multiplication by a continuous piece-wise analytic scalar function $a(k)$

$$(Af)(k) = a(k)f(k). \tag{5.7.2}$$

In other words, for any of the three components of the spectrum to be able to appear in $\sigma(H)$, it should appear in one of the multiplication operators (5.7.2) from Theorem 5.6.3. Thus, we concentrate on the operator A in 5.7.2 for now.

Theorem 5.7.1 *The following statements hold:*

(1) *The singular continuous spectrum of the operator A is empty:*

$$\sigma_{\text{sc}}(A) = \emptyset. \tag{5.7.3}$$

(2) *The pure-point spectrum consists of all values λ such that the λ-level set of $a(k)$ has positive measure. In particular, if there is no positive measure "flat piece" $a = \text{constant}$, then $\sigma_{\text{pp}}(A) = \emptyset$ and thus the whole spectrum is absolutely continuous:*

$$\sigma(A) = \sigma_{\text{ac}}(A).$$

(3) *For each $\lambda \in \sigma_{\text{pp}}(A)$, the function $a(k)$ is constant on an open set.*

The sketch of the proof is provided in Appendix F.

We thus get the following result:

Theorem 5.7.2 *For any periodic self-adjoint elliptic operator H, one has the following:*

(1) *The singular continuous spectrum is empty: $\sigma_{\text{sc}}(H) = \emptyset$.*
(2) *The pure-point spectrum consists of all values λ such that the λ-level set of some of $\lambda_j(k)$ has positive measure.*

(3) *If the pure-point spectrum is non-empty, then at least one of the band functions* λ_j *is constant on an open set.*
(4) *The pure-point spectrum consists of all values* λ_0*, such the complex Bloch variety* $B_{H,\mathbb{C}}$ *contains the flat component* $\lambda = \lambda_0$.
(5) *The ac-spectrum is always present.*

Proof The first three statements are corollaries of Theorems 5.6.3 and 5.7.1. The last two follow from the same two theorems and Theorem 4.6.12. □

Corollary 5.7.3 *Under the same conditions,*

$$\sigma(H) = \sigma_{\mathrm{ac}}(H) \cup \sigma_{\mathrm{pp}}(H). \tag{5.7.4}$$

In other words, unless the operator H has a true eigenfunction (a bound state), its spectrum is absolutely continuous.

Thus, the remaining (and highly nontrivial) question is:

Is $\sigma_{\mathrm{pp}}(H) = \emptyset$ for any periodic self-adjoint elliptic second-order operator[6] H?
Here "any" assumes sufficiently nice coefficients, for example, the ones guaranteeing at least a weak uniqueness of continuation of solution.

Only some partial (both positive and negative) answers are known, which we discuss in the following section.

5.7.1 Absence/Presence of σ_{pp}

Now the question remains whether both pure-point and absolutely continuous parts of the spectrum can be present. We prefer to ask the question of presence of the pure-point spectrum as of existence of L_2-eigenfunctions, since then it can be formulated also in the non-self-adjoint situation.

5.7.1.1 Flat Components and the Pure-Point Spectrum

Here we show the relation of the question of existence of the pure-point spectrum to the geometry of the Bloch variety.

Theorem 5.7.4 (Thomas, 1973) *The following two statements about a periodic elliptic operator H are equivalent:*

(1) *The Bloch variety* B_H *contains the flat component* $\{(k, \lambda) \mid k \in \mathbb{C}^n\}$.
(2) *For* $\lambda \in \mathbb{C}$ *there exists an* L_2*-solution of the equation* $Hu = \lambda u$ *(in the self-adjoint case,* $\lambda \in \sigma_{\mathrm{pp}}(H)$*).*

[6] See Remark 5.7.5 to see that things can break down for operators of higher orders.

Proof (1) $\Rightarrow$ (2) According to this assumption, the kernel of the operator $(H(k) - \lambda I)$ is not zero for all values of k. Pick a real value $k_0 \in \mathbb{R}^n$ at which the dimension of this kernel is minimal among nearby real values of the quasi-momentum. Then, as it has been explained before, these kernels form a trivial analytic vector bundle near k_0. Picking a compactly supported smooth section $\widehat{u}(k)$ of this bundle and applying to it the inverse Floquet transform, one obtains an L_2-solution of $(H - \lambda I)u = 0$.

(2) $\Rightarrow$ (1) Let $u \in L_2$ be a nonzero solution of

$$(H - \lambda I)u = 0. \tag{5.7.5}$$

Floquet transform provides an L_2-vector function $\widehat{u}(k)$ of k whose values belong to $\mathrm{Ker}(H(k) - \lambda I)$. Since the support of an L_2- function has a positive measure, the conclusion is that the Fermi surface $F_{\lambda,H}$ intersect the space of real quasi-momenta over a set of positive measure. Then Theorem 4.6.11 finishes the proof. □

5.7.1.2 Some Counterexamples and Relation to Uniqueness of Continuation

The first thing (which sometimes had been missed) to notice is the following:

Remark 5.7.5 Absolute continuity of the spectrum **fails for some periodic elliptic operators of order higher than** 2.

The answer happens to be related to validity of the **weak uniqueness of continuation property**[7] for the given operator. For instance, there is a known example (Pliś, 1960) of a fourth-order elliptic operator in $\mathbb{R}^3$ of the form

$$P(x, D) = \psi(x)R(x, D) + (1 - \psi(x))\Delta^2, \tag{5.7.6}$$

which has a nontrivial compactly supported function $u(x)$ in its kernel:

$$P(x, D)u(x) = 0.$$

Here $\psi(x) \in C_0^\infty(\mathbb{R}^n)$ is a smooth cut-off function and $R(x, D)$ is an elliptic fourth-order operator with smooth coefficients. Now one can find a large cube that contains both the support of ψ and the support of u. Using this cube as the fundamental domain, one can extend $P(x, D)$ to a periodic elliptic operator $H(x, D)$ with smooth coefficients in $\mathbb{R}^n$. This new operator still satisfies the same equation $H(x, D)u(x) = 0$, and thus $0 \in \sigma_{\mathrm{pp}}(H)$.

This relation to the uniqueness of continuation theorems does not seem to be accidental. For instance, an example of a scalar second-order periodic elliptic operator with non-empty pure-point spectrum was constructed in Filonov (2001). In this example, the leading (i.e., in the second-order terms) coefficients fall just a notch below of what is needed for the weak uniqueness of continuation to hold.

Another, although less direct, evidence of this is that one can prove that existence

[7] That is, the absence of nonzero solutions vanishing on an open sub-set.

of a "true" (i.e., L^2) eigenfunction implies existence of a much faster (exponentially, or even super-exponentially) decaying one, which is much less believable and should violate some "**unique continuation at infinity**" result.[8] Let us make this more precise in the following statements:

Theorem 5.7.6 *Let H be an elliptic periodic operator (of an arbitrary order). If the equation $Hu = 0$ has a nontrivial L^2-solution u, it also has a solution $v \neq 0$ such that*

$$|v(x)| \leq C_a e^{-a|x|} \tag{5.7.7}$$

for any $a > 0$.

Proof Suppose that $\lambda_0 \in \sigma_{pp}(H)$. Incorporating λ_0 into the operator, we can assume for simplicity that $\lambda_0 = 0$. Then, according to Theorem 5.7.4, for any point $(k, 0) \in \mathbb{C}^{n+1}$, the equation $H(k)u = 0$ on the torus has a nonzero solution. Let us move to the Floquet multipliers $z = e^{ik}$ instead of quasi-momenta k. Then, Theorem C.5 implies existence of analytic function $w(z)$ on $(\mathbb{C} \setminus \{0\})^n$ with values in smooth functions on the torus, such that $H(z)w(z) = 0$ for all z. We now apply the second statement of Theorem 3.4.16 in case when $a = \infty$ to conclude that w is the Floquet transform of an exponentially decaying function $v(x)$ in $\mathbb{R}^n$. This function v clearly satisfies all claims of the Theorem. □

One can achieve even stronger decay:

Theorem 5.7.7 *Let H be an elliptic periodic operator in $\mathbb{R}^n$ (of an arbitrary order) such that equation $Hu = 0$ has a nontrivial L^2-solution u. Then*

(1) *there exists a nontrivial solution $v \neq 0$ that decays super-exponentially. That is,*

$$|u(x)| \leq Ce^{-|x|^\gamma}, \tag{5.7.8}$$

with some $\gamma > 1$;

(2) *if the equation $Hu = 0$ has the weak uniqueness of continuation property, it has a solution $v \neq 0$ such that*

$$|v(x)| \leq Ce^{-C|x|^{n/(n-1)}}. \tag{5.7.9}$$

We will not provide the more technical proof of this theorem here (see Theorem 4.1.6 in Kuchment, 1993).

As we have already mentioned, the following is known as **Landis' conjecture**:

Conjecture 5.7.8 *Nonzero super-exponentially decaying solution, as in Theorem 5.7.7, do not exist (they violate some expected "unique continuity at infinity" result).*

[8] The absence of such fast decaying solutions is known to be the **Landis' conjecture**. See Appendix E.

Regretfully, the only such result known to the author, which would apply to periodic Schrödinger operators (Froese et al., 1983; Meshkov, 1991), deals with these with electric potentials only and is not strong enough to lead to a new absolute continuity theorems. For example, it does not allow variable coefficients in the terms of the first and second orders. Although there is a renewed activity in the direction of proving the Landis' conjecture (see, e.g., Froese et al., 1983; Kenig et al., 2015; 2019; Logunov and Malinnikova, 2018; Meshkov, 1991, and Appendix E for more details), it has not led to new absolute continuity results yet.

Another indication on the relation with the uniqueness of continuation comes from equations on discrete or quantum graphs, for which the uniqueness of continuation does not hold. And sure enough, one can find compactly supported eigenfunctions for second-order elliptic periodic operators on such graphs (see Berkolaiko and Kuchment, 2013, references therein, and Fig. 5.4).

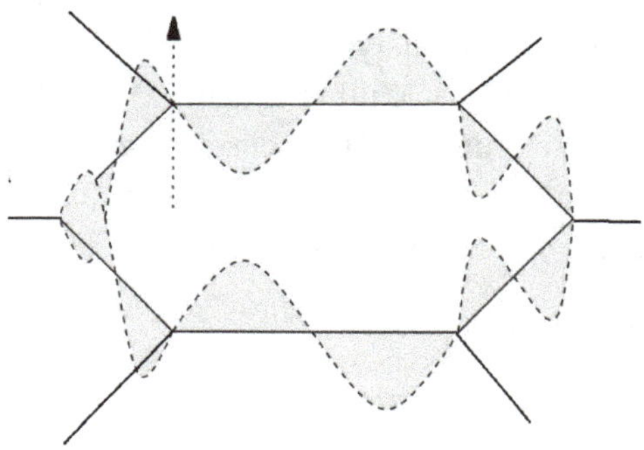

Figure 5.4 A compactly supported eigenfunction on a quantum graph.

Let us see now what happens when there is uniqueness of continuation, for example, for "nice" second-order operators (Mizohata, 1973).

Second-Order Operators

Using the last statement of Theorem 5.7.2 and Theorem 4.6.16 that prohibit existence of flat components in the Bloch variety for the Schrödinger operator $-\Delta+V(x)$, one reaches the following conclusion (the famous theorem of L. Thomas):

Theorem 5.7.9 (Thomas, 1973) *If the periodic potential V is bounded (and not necessarily real), then the spectrum of the Schrödinger operator with this potential does not have true eigenvalues. (That is, in the self-adjoint case, the spectrum is absolutely continuous.)*

As it was the case with Theorem 4.6.16, here the condition of boundedness of the periodic potential is a significantly stronger than necessary (see, e.g., Section VIII.16 of Reed and Simon, 1972).

Remark 5.7.10 One should not get over-excited about dropping conditions on the coefficients in the highest order terms, since when they are going below those guaranteeing weak uniqueness of continuation property, point spectrum can arise, as it was shown in Filonov (2001).

Since the famous work of Thomas, practically all proofs of the absolute continuity of the spectra of periodic elliptic operators followed the same scheme that we have used: proving the absence of flat components in the dispersion relation (Bloch variety). Due to the hard technical work of many researchers, absolute continuity of the spectrum has been proven for many (**but still not all**) second-order scalar elliptic periodic operators, as well as periodic Pauli, Dirac, elasticity theory, and in some cases Maxwell operators, for example, see Birman and Suslina (1999b, 2002, 2000, 1999a); Birman et al. (2000a); Danilov (1990, 1995, 1999, 2000b,a, 2003b,a, 2005, 2009, 2010, 2011); Filonov and Sobolev (2004); Friedlander (2002); Gérard and Nier (1998a,b); Hempel and Herbst (1995), Karpeshina and Lee (2008); Krupchyk and Uhlmann (2017); Kuchment (1993); Kuchment and Levendorskiî (2002); Reed and Simon (1972); Shen (2001); Sobolev (1999); Thomas (1973); Tikhomirov and Filonov (2004); and references therein. However, the following conjecture still remains unproven, even under assumption of infinite differentiability of the coefficients of the operator:

Conjecture 5.7.11

(1) *The spectrum of any self-adjoint second-order periodic elliptic scalar operator with "nice" (e.g., smooth) coefficients is absolutely continuous.*
(2) *In the non-self-adjoint case (when classification into three types of spectra is not applicable), there are no eigenvalues in the spectrum.*

In all works known to the author, except Friedlander (2002) and few papers based upon it (e.g., Tikhomirov and Filonov, 2004), the proof goes analogously to the one of Theorem 4.6.16: dominating in the Fourier domain the lower order terms by the principal part. This approach gets much harder in the presence of variable coefficients in the first-order (magnetic potential) terms (see Sobolev, 1999) and it does not look at all feasible to treat variable coefficients in the principal part of the operator. The least technical approach, which beautifully avoids these domination estimates and thus allows for variable coefficients in the second-order terms, was introduced in Friedlander (2002), where a breakthrough was achieved in terms of treating more general operators. We present this result here. In order to do so, let us describe first the type of the operator we will deal with:

$$H = -\sum_{p,l=1}^{n} \frac{\partial}{\partial x_p} g_{pl}(x) \frac{\partial}{\partial x_l} + \frac{1}{i} \sum_{l=1}^{n} \left(a_l(x) \frac{\partial}{\partial x_l} + \frac{\partial}{\partial x_l} a_l(x) \right) + V(x). \quad (5.7.10)$$

We assume here that all coefficients g_{pl}, a_l, and V are smooth, real valued, and $\mathbb{Z}^n$-periodic, where $\mathbb{Z}^n$ is acting by shifts in $\mathbb{R}^n$ (the restriction on the group is so far irrelevant). It is assumed that $g_{pl} = g_{lp}$. So far this makes H a general elliptic periodic self-adjoint operator of second order with smooth coefficients.

Theorem 5.7.12 *Let H be as stated previously. Assume additionally that the operator H is invariant with respect to the symmetry $(x_1, x_2, \ldots, x_n) \mapsto (-x_1, x_2, \ldots, x_n)$. Then $\sigma_{\text{pp}}(H) = \emptyset$, and thus the spectrum of H is absolutely continuous.*

Remark 5.7.13 Let us make a comment concerning the symmetry condition imposed.

(1) The symmetry condition in terms of the coefficients can be stipulated as follows (denoting $(x_2, \ldots, x_n) = x'$):

$$\begin{cases} g_{11}(-x_1, x') = g_{11}(x_1, x') \\ g_{pl}(-x_1, x') = g_{pl}(x_1, x'), a_l(-x_1, x') = a_l(x_1, x') \text{ for } p, l \geq 2 \\ g_{1l}(-x_1, x') = -g_{1l}(x_1, x'), a_1(-x_1, x') = -a_1(x_1, x'), l \geq 2 \\ V(-x_1, x') = V(x_1, x'). \end{cases} \quad (5.7.11)$$

(2) The symmetry condition seems to be irrelevant for the absolute continuity statement (or at least strange). However, so far no one has been able to remove it.

The author does not want to deprive the reader of the pleasure of reading the beautiful concise original proof in Friedlander (2002), based upon the use of the Dirichlet-to-Neumann operator.

There are also quite a few, often more demanding, results on absolute continuity for Maxwell operators, operators in periodic waveguides, on periodic systems of curves and surfaces, etc. (Bentosela et al., 2003; Derguzov, 1972b; Exner and Frank, 2007; Filonov and Klopp, 2004a,b, 2005; Friedlander, 2003; Kachkovskiĭ and Filonov, 2009, 2010; Kuchment, 1982; Shterenberg and Suslina, 2001; Sobolev and Walthoe, 2002; Suslina, 2001, 2002), which we plan to address in Kuchment (2027).

As we have mentioned in Section 4.6.6, the presence of a flat component implies reducibility of the Bloch variety, a topic discussed there in detail.

5.8 Detecting the Spectrum: Bloch–Shnol'-Type Theorems

The standard linear algebra definition of the spectrum describes it as the set of all eigenvalues of a matrix, that is, in terms of existence of eigenvectors. In the setting of infinite dimensional self-adjoint operators, however, this defines only the pure-point spectrum (which, in most cases of our interest is empty). On the other hand, the PDE $Hu = \lambda u$ might have a nontrivial solution u that does not belong to $L^2(\mathbb{R}^n)$ and thus is not a "true" eigenfunction (it is called **generalized eigenfunction**). In fact, for PDEs of the class we discuss, there is a way to have a complete (in a specific sense) system of such generalized eigenfunctions and spectral expansion into them (the famous and often very useful **generalized eigenfunctions expansion**). Moreover, one can often guarantee some limits on the growth of such solutions (e.g., polynomial or at least sub-exponential). The reader is referred to Berezanskii (1968); Berezin and Shubin (1991); Gel′fand and Kostyučenko (1955); Gel′fand and Shilov (2016); and Krein et al. (1972) for details and the results of the following nature:

Meta-Theorem 5.8.1 *Unlike the "true" L^2-eigenfunctions, generalized eigenfunctions of an appropriately limited growth exist almost everywhere in $\sigma(H)$ with respect to the spectral measure.*

Moreover, one can create spectral decomposition for H using these.

Now the following question arises: **Does a converse statement hold: existence for a given λ of a generalized eigenfunction of suitably controlled growth imply that $\lambda \in \sigma(H)$?**

The answer is positive and is supplied by the so-called **Shnol' theorem** (Cycon et al., 1987; Glazman, 1966; Shnol', 1957). In the periodic case, a version of this is known in physics as the following **Bloch theorem**:

Theorem 5.8.2 *If the equation*

$$-\Delta u + V(x)u = \lambda u$$

in $\mathbb{R}^n$ with a real locally square-summable periodic potential V has a nontrivial bounded generalized eigenfunction, then it has a nontrivial bounded Bloch solution $e^{ik\cdot x}p(x)$, and thus $\lambda \in \sigma(H)$.

We are not going to prove this statement, since we will obtain a much stronger result.

Shnol'-type results, for example, Cycon et al. (1987); Glazman (1966); Kuchment (1982, 1993); Shnol' (1957); and Shubin (1990), say that instead of requiring a bounded solution, it is sufficient to have a sub-exponentially growing solution, or in fact, a family of solutions $u_\epsilon(x)$ such that $|u_\epsilon(x)| \le C_\epsilon e^{\epsilon|x|}$ for a sequence of values of $\epsilon > 0$ that tend to zero.

Another, stronger, version of the Shnol' theorem says that if there exists a generalized eigenfunction of a given exponential growth $|u(x)| \leq Ce^{a|x|}$, then one can estimate from above the distance of λ to the spectrum of H (with the distance going to zero when $a \to 0$).

Such theorems can be proven under conditions weaker than periodicity. We, however, present here a useful version of the Shnol' theorem that is specifically adopted to the periodic case.

Theorem 5.8.3 (Kuchment, 1993) *Let $H(x, D)$ be a periodic elliptic operator in $\mathbb{R}^n$. If for some spectral value λ operator H has a generalized eigenfunction with the estimate*

$$|u(x)| \leq Ce^{a \sum |x_j|}, \tag{5.8.1}$$

then it also has a Bloch generalized eigenfunction with a quasi-momentum $k \in \mathbb{C}^n$, such that

$$|\operatorname{Im} k_j| \leq a, \qquad j = 1, \ldots, n. \tag{5.8.2}$$

In other words, this Bloch solution also satisfies (5.8.1).

The proof can be found in Kuchment (1993, Theorems 4.1.5 and 4.1.6).

Corollary 5.8.4 *Suppose that for some λ there exists a sequence of generalized eigenfunctions $u_n(x)$ such that $|u_n(x)| \leq C_n e^{\epsilon_n |x|}$, where $\epsilon_n \to 0$ when $n \to \infty$. Then $\lambda \in \sigma(H)$.*

This, in particular, is satisfied if there exists a generalized eigenfunction of a sub-exponential growth.

Indeed, this statement follows from Theorems 4.6.11, 5.8.3, and closedness of the Fermi surface.

Remark 5.8.5

- Notice that self-adjointness of the operator is not required in these results.
- The way these results are formulated, they would not apply on Riemannian manifolds, if the volume of a ball B_r grows faster than polynomial with $r \to \infty$. This is, however, an artifact of the formulation. We discuss in the remarks Section 5.14 how the statement can be reformulated to be applicable without any volume growth restrictions.

5.9 Density of States

An important object in spectral theory and solid-state physics is the so-called **density of states (DOS)** of a self-adjoint periodic elliptic operator H. Here is the idea: consider an "appropriate" sequence of expanding when $N \to \infty$ domains $V_N \subset \mathbb{R}^n$, eventually covering the whole space (see, e.g., Remark 5.2.5). After imposing "appropriate" boundary conditions on these domains, one can consider the counting function of eigenvalues λ_j of the resulting operator with discrete spectrum:

$$N(\lambda) := \#\{\lambda_j < \lambda\}. \tag{5.9.1}$$

Now one normalizes this function by the volume of the domain V_N and takes the limit when $N \to \infty$:

Definition 5.9.1 The **integrated density of states (IDS)** of the operator H is

$$N(\lambda) := \lim_{N\to\infty} \frac{1}{|V_N|} \#\{\lambda_j < \lambda\}. \tag{5.9.2}$$

One certainly wonders whether the limit exists and is independent of the choice of the expanding domains V_N and boundary conditions imposed. Under appropriate conditions on those, the answer is a "yes" to both questions (see Shubin, 1979, where this is done even for almost periodic coefficients; and Adachi and Sunada, 1993, for a more general discussion).

Without listing the conditions, we just acknowledge that, assuming that the lattice is transformed to $\mathbb{Z}^n$, this is true when one picks V_N as the sequence of cubes $\{x \in \mathbb{R}^n \,||x_j| \le N,\ j = 1, \dots, n\}$ for $N \in \mathbb{Z}^+$ and imposes periodic boundary conditions on those (the Born–Karman conditions of Section 5.2.1).

A simple calculation shows that the IDS can be described in terms of the dispersion relation $\lambda_j(k)$ as follows:

Theorem 5.9.2

$$N(\lambda) = \sum_j \mu\{k \in \mathbb{T}^* \,|\, \lambda_j(k) < \lambda\}, \tag{5.9.3}$$

where μ is the previously defined Haar measure on the torus $\mathbb{T}^*$.

Due to the analytic nature of the band functions studied before, one obtains the following corollary:

Corollary 5.9.3

(1) *The IDS is a piecewise analytic function of λ.*
(2) *Unless a flat band function $\lambda =$ constant is present, the IDS is continuous.*
(3) *Singularities of the IDS can arise if λ is a singular or critical value of the dispersion relation (**Van Hove singularities**, see Section 5.10).*

Definition 5.9.4 The **DOS** is the Radon–Nikodym derivative of the IDS with respect to the Lebesgue measure:

$$g(\lambda) := \frac{dN}{d\lambda}. \tag{5.9.4}$$

Figure 5.5 illustrates these notions:

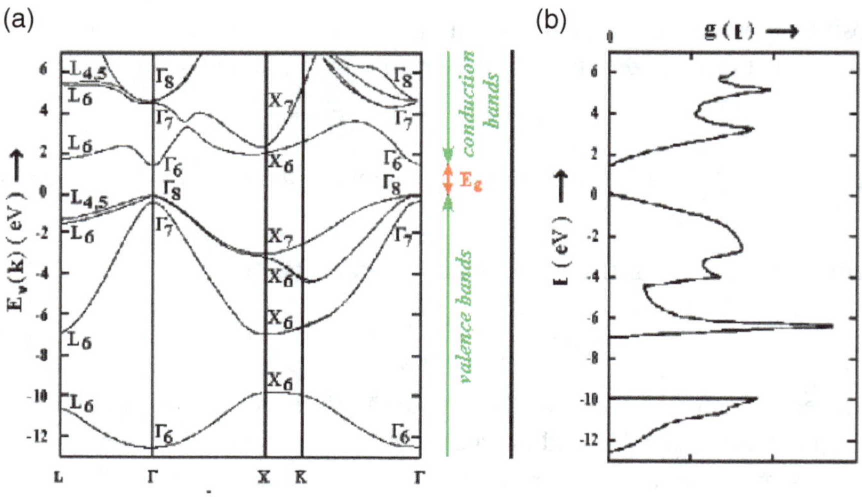

Figure 5.5 (a) Dispersion relation. (b) The corresponding density of states.

The following explicit formula in terms of the band functions λ_j follows from (5.9.2).

Theorem 5.9.5

$$g(\lambda) = (2\pi)^{-n} \sum_j \int_{\lambda_j=\lambda} \frac{ds}{|\nabla_k \lambda_j|}. \tag{5.9.5}$$

Remark 5.9.6 This calculation shows, as we will see later, the absence of a singular continuous spectrum.

5.10 Van Hove Singularities

The formula (5.9.5) for the DOS indicates that the DOS might have singularities, called **Van Hove singularities** (Van Hove, 1953) of the following types:

(1) at critical points (extrema or saddle points) of a single (and thus analytic) band function $\lambda_j(k)$, or
(2) at intersections of bands (Figure 5.6).

Due to the related presence of high DOS, Van Hove singularities are known to trigger important and often unusual physical effects, for example, superconductivity (see essentially any solid state physics text, e.g., Ashcroft and Mermin, 1976; Bassani and Pastori Parravicini, 1975; Callaway, 1964; Ziman, 1964, as well as the nice survey

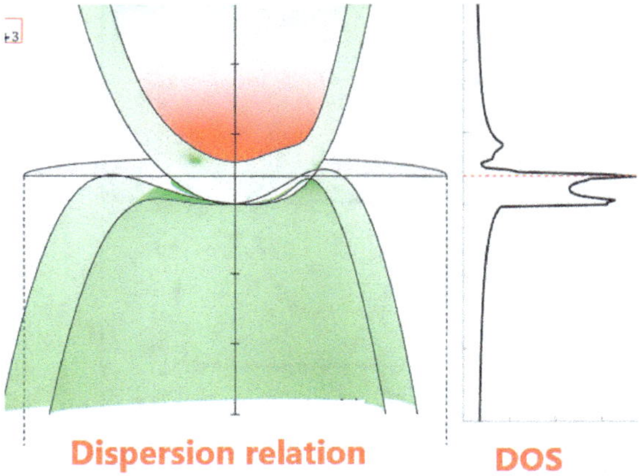

Figure 5.6 Van Hove singularities.

article Dugdale, 2016). They have been studied mostly for Schrödinger operators with periodic electric potentials. Van Hove singularities also appear when one studies limiting absorption for periodic operators, see, for example, Gérard (1990) and Section 5.12.

One can expect that same order of vanishing of the gradient $\nabla_k \lambda_j$ creates a less severe singularity in the DOS in higher dimensions: in 1D, DOS diverges at these points, while in 2D one normally sees square-root-type singularities (see Figure 5.7). In 3D, Van Hove singularities seem to be infrequent, and their discoveries in physical situations usually attract attentions, see, for example, the (rather random) collection of articles Dugdale (2016); Hu et al. (1998); Igoshev (2022); and Wu et al. (2024).

5.11 Spectral Asymptotics

Another important and active area of research is high-energy spectral asymptotics (e.g., asymptotic of DOS) for periodic elliptic operators. In this section, we provide just a pointer to the history and main results in this large and striving area. You can find the complete history and the proof of the main conjecture in Morozov et al. (2014).

Consider the Schrödinger operator

$$H = -\Delta + V.$$

With smooth and periodic[9] operator in $\mathbb{R}^d$. Let also $N(\lambda)$ be the corresponding IDS

[9] The existing results have been largely extended to (at least generic) **almost-periodic potentials, as well as to pseudo-differential** operators (Morozov et al., 2014).

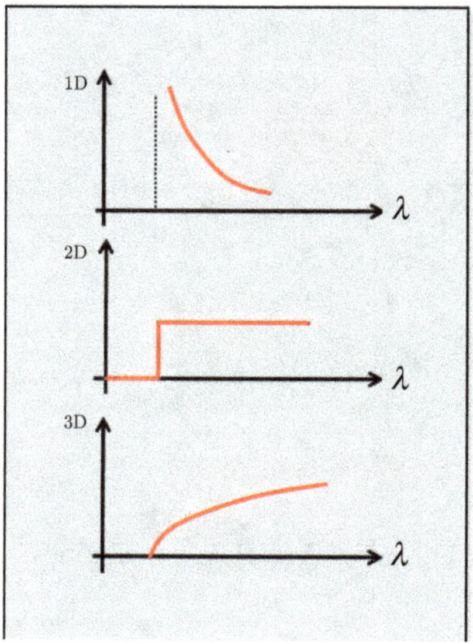

Figure 5.7 Examples of Van Hove singularities in different dimensions.

(see Section 5.9). For the free operator $-\Delta$, one can compute explicitly the IDS N_0:

$$N_0(\lambda) = C_d \lambda^{d/2},$$

with explicit known dimension-dependent constant C_d (see Morozov et al., 2014).

A long-standing conjecture is:

Conjecture 5.11.1 *In general, the following asymptotic expansion holds:*

$$N(\lambda) = \lambda^{d/2}\left(C_d + \sum_{j=1}^{m} e_j \lambda^{-j}\right) + r_m(\lambda), \tag{5.11.1}$$

where $r_m(\lambda) = o(\lambda^{d/2-m})$.

Assuming validity of this conjecture, the coefficients e_j can be found explicitly using heat invariants (see Hitrik and Polterovich, 2003).

After a long history, which started with the 1D work of Shenk and Shubin (1986, 1987) and is well-described in Morozov et al. (2014), the previous conjecture, as well as its far-reaching extensions to operators of higher order and stronger (including magnetic) perturbations was proved in Morozov et al. (2014).

One cannot avoid mentioning the magnificent treatise on general theory of spectral asymptotics of Ivrii (1998, 2019b,c,d,e,f).

One can find the newest developments, the history, and references in Ivrii (2021).

We should also mention semi-classical spectral asymptotics (see, e.g., Guillot et al., 1988).

5.12 Limiting Absorption

Here we briefly remind the reader the notion of the **limiting absorption principle** and some results available in the periodic Schrödinger operators case. One can find historic accounts and nice explanations of this notion (as well as of closely related radiation and limiting amplitude principles) in many publications, for instance, in Tao (2011) and Vaĭnberg (1966).

Let A be a self-adjoint operator in Hilbert[10] space $\mathcal{H}$. Then its spectrum belongs to the real line $\mathbb{R}$ and its resolvent $R(\lambda) = (A - \lambda I)^{-1}$ exists and is analytic in the complex domain $\mathbb{C}\backslash\mathbb{R}$. If the spectrum of A is discrete, then the resolvent analytically extends through the real line, avoiding only the eigenvalues. The limiting absorption principle (see the explanation of this name in the papers listed above) deals with what happens with the resolvent, as a bounded operator, when there is essential spectrum and we approach it from above or below. Are there limits? It seems to be unlikely, and not hard to prove impossible, when we approach the points of the spectrum. It might be possible, however, if we weaken the condition on the resolvent of being a bounded operator in $\mathcal{H}$. Namely, we can require only that the limit of the resolvent is continuous as acting from a smaller dense subspace $\mathcal{H}_+ \subset \mathcal{H}$ with the range in a larger space $\mathcal{H}_-$, $\mathcal{H} \subset \mathcal{H}_-$.[11] The standard choice when $\mathcal{H} = L_2(\mathbb{R}^n)$ is that the smaller space consists of $L_{2,\mathrm{loc}}$-functions with a given exponential (or algebraic) decay at infinity, while the larger space allows the corresponding exponential (or algebraic) growths. If this relaxation of requirements leads to the existence of the limits of the resolvent from above/below the spectrum, we say that the **limiting absorption principle holds**. There are plenty of results on validity of the limiting absorption for Laplace and some more general differential operators (see again Tao, 2011; Vaĭnberg, 1966 and the references therein).

In Murata and Tsuchida (2006), the limiting absorption principle is established near the bottom of the spectrum for Schrödinger operators with periodic electric potentials. This is a nontrivial technical result, albeit underlying intuition is clear: the dispersion relation near the bottom of the spectrum has parabolic shape and thus one expects behavior analogous to the case of Laplacian.

Gérard (1990) contains the powerful result on validity of the limiting absorption principle for periodic Schrödinger operators, as long as one avoids Van Hove singularities.

10 Some versions for Banach spaces also exist.

11 This construction is called a **rigged Hilbert space**, or **Gelfand's triple** (Arlinskii et al., 2011; Berezanskii, 1968; Berezansky et al., 1996; Gel′fand and Vilenkin, 2016).

Remark 5.12.1

(1) The technique used in Murata and Tsuchida (2006), as well as, most probably, the result itself, might not work for a general periodic magnetic potential, due to possible more complex behavior of the dispersion relation at the bottom of the spectrum, as for instance the result of Shterenberg (2004) shows.
(2) The author conjectures that similar results should work near any nondegenerate spectral band edge and any periodic elliptic operator of the second order. See, for instance, the asymptotics of the Green's function near the band edges (Kha, 2018; Kha and Kuchment, 2021; Kha et al., 2017).

5.13 Inverse Spectral Problems

There are several inverse spectral problems of interest for periodic operators. One can use different types of spectral data to try to recover some features of the operator (e.g., potentials in Schrödinger case). These could be spectral gaps, a spectrum (e.g., the spectrum on periodic functions), the whole Bloch variety, a Fermi surface, etc. All of these questions happen to be rather hard to resolve in dimensions higher than 1. In this section, we survey some existing results and outstanding conjectures.

One of possible formulations of an inverse problem is suggested by the classical Borg uniqueness theorem for 1D Schrödinger operators:

Theorem 5.13.1 (Borg, 1946) *If the spectrum of the self-adjoint* 1D *Schrödinger operator*

$$H = -\frac{d^2}{dx^2} + V(x) \tag{5.13.1}$$

with a periodic potential has no gaps, then the potential is constant. The converse statement also trivially holds.

As we saw in Section 5.4, this is no more true for Schrödinger operators $-\Delta + V(x)$ with periodic potentials in higher dimensions. Indeed, any sufficiently small potential does not produce any gaps when $n = 2, 3$. Thus, the straightforward restatement of Borg's uniqueness theorem fails. The reason is that gaps themselves do not seem to carry much of tangible information about the operator in dimension higher than one. This will be seen as a big disappointment, when one discusses the problems of photonic crystal theory in Kuchment (2027) (see also Kuchment, 2001). However, there is an equivalent (in 1D) reformulation of the 1D Borg uniqueness theorem that appears to hold in higher dimensions as well.

Theorem 5.13.2 *Suppose there exists an entire function $\lambda(k)$ such that its graph belongs to the Bloch variety B_H of the* 1D *Schrödinger operator* (5.13.1). *Then the potential is constant.*

The converse statement also (trivially) holds.

It is likely that the analog of this version of Borg's theorem holds in any dimension:

Conjecture 5.13.3 *If the Bloch variety of the Schrödinger operator* $-\Delta + V(x)$ *in* $\mathbb{R}^n$ *with a real periodic potential contains an entire branch* $\lambda(k)$*, then the potential is constant.*

This statement was proven in the 2D case:

Theorem 5.13.4 *If* $H = -\Delta + V(x)$ *in* $\mathbb{R}^2$*,* $V \in L^2_{loc}(\mathbb{R}^2)$ *is real and periodic, and the Bloch variety* B_H *contains an entire branch* $\lambda(k)$*, then the potential is constant.*

We will not provide the very technical proof of this theorem, which can be found in Knörrer and Trubowitz (1990).

Some works address what properties are shared between isospectral potentials. One question of this type is whether if a potential is split, is any isospectral potential also split? We consider potentials in $\mathbb{R}^n$ that are **split** into a sum of potentials depending only on parts of the variables, for example:

$$V(x) = V_1(x_1) + V_2(x_2) + V_3(x_3), \tag{5.13.2}$$

or a more general class of split potentials:

$$V(x) = V_1(x_1) + V_2(x_2, x_3). \tag{5.13.3}$$

Theorem 5.13.5 (Bättig et al., 1991) *Let* $H = -\Delta + V(x)$ *with the potential* $V(x) \in L^2_{loc}(\mathbb{R}^3)$ *being real and* $\mathbb{Z}^3$*-periodic. If for some* $\lambda \in \sigma(H)$*, the real Fermi surface* $F_{\mathbb{R},\lambda}(H)$ *coincides with the one corresponding to a split potential of the type* (5.13.2) *or* (5.13.3)*, then* $V(x)$ *is also split.*

Several theorems of this nature can also be found in the wonderful series of works Eskin et al. (1984a,c).

5.14 Comments and References

- **Wave packets and group velocity:** The name "dispersion relation" for the band functions $\lambda(k)$ resonates with the corresponding notion in the theory of dispersive waves (see, e.g., Brillouin, 1960). This brings out an interest in studying propagation of **wave packets** of Floquet–Bloch eigenfunctions. Certainly, this concerns evolution equations, which we discuss in Kuchment (2027). Let us still say a few words about it. Let $\psi(\lambda, k)(x)$ be such an eigenfunction depending analytically on the point (λ, k) in a neighborhood U of the point (λ_0, k_0) of the spectral band $\lambda_j(k)$. Let also consider a small "bump" mollifier $\phi(\lambda, k)$ at this point, then the **wave packet** is

$$f(x) := \int_U \psi(\lambda, k)(x)\phi(\lambda, k)\, dk\, d\lambda\,. \tag{5.14.1}$$

An important notion of the **group velocity** of this packet propagation (essentially, the velocity of the propagation of the shape). It is determined by the k- component of the gradient to the dispersion relation at the center (λ_0, k_0) of the packet. For instance, at critical points of dispersion relation, the group velocity vanishes, and thus the packet "freezes."[12] Certainly, dynamics eventually destroys the packet – it "spreads" – but this approximation is valid for some time that can be estimated (see, e.g., Fefferman and Weinstein, 2014; Molchanov and Vainberg, 2004, 2005, 2006).

- **Section 5.1:** Theorem 5.1.1 is a particular case of more general result that holds for **almost periodic operators** (Shubin, 1979).
- **Section 5.2:** The **Born–Karman approximation** is frequently used for computing spectra of periodic operators. However, Theorem 5.2.6 is more than just the claim about Hausdorff convergence of spectra or of the DOS (see discussion of the latter in Section 5.9). Indeed, such claims hold also for other "reasonable" choices of boundary conditions on Q_N (e.g., Dirichlet) or other choices of large domains instead of large cubes (Shubin, 1979). In those cases, however, one can encounter **spurious spectra** appearing outside $\sigma(H)$ (i.e., in the gaps), but disappearing in the limit when $N \to \infty$ (e.g., Levitin et al., 2010). In the Born–Karman case, though, this does not happen. Indeed, clearly $\sigma(H^{(N)}) \subset \sigma(H)$ for any N. Thus, in the Born–Karman procedure, the spectrum of H is approximated "from within" and spectral gaps are not "polluted."
- **Section 5.4 Spectral gaps and the Bethe–Sommerfeld conjecture:**
  - As we have seen, both in 1D and in the higher dimensions, periodic Schrödinger (and more general elliptic) operators have band-gap structure of their spectra. We have realized, though, that there are significant differences between ODE and PDE cases. We compare them in Table 5.1.
  - By now, a century after it was formulated and after several decades of efforts, the **Bethe–Sommerfeld conjecture** has been proven in its full generality.[13] The story started with Dahlberg and Trubowitz (1982); Popov and Skriganov (1981); and Skriganov (1985b) for the most interesting physical cases $n = 2$ and $n = 3$. The approach used was, as described in Section 5.4, through estimating the number of overlapping bands. Theorem 5.4.6 shows that the

[12] One can, for instance, "slow down" light, see, for example, Figotin and Vitebskiy (2006); Heebner et al. (2002); Kuchment and Kunyansky (1999); Smyshlyaev and Kuchment (2007); Yariv et al. (1999).

[13] The history of its proof is rather complicated, and the author thanks the referees for their help.

Table 5.1 *A comparison of ODE and PDE cases*

	ODEs, $n = 1$	PDEs, ne
Band overlaps	None	Frequent
Gaps	Generically all open	Many gapless potentials
No gaps	Constant potential	All small potentials for $n = 2, 3$
Free operator bands	Touch to cover $\mathbb{R}^+$	Overlap to cover $\mathbb{R}^+$

issue should become much harder for $n > 3$, and it indeed does. For $n \geq 4$, the conjecture was established for the case of rational lattices in Skriganov (1985a); Skriganov and Sobolev (2005); and for arbitrary lattices in Helffer and Mohamed (1998). Finally, in the remarkable work of Parnovski (2008), the proof was provided in any dimension $n \geq 2$ and arbitrary lattice, under some smoothness conditions on the (electric) potential.

- Predictably, the case of presence of **magnetic potential** turned out to be even harder. The result was established in the 2D case in Karpeshina (1997) and Mohamed (1997). For the most general case of stronger periodic perturbation, see Parnovski and Sobolev (2010). A major advance both in techniques and results (allowing not only lower-order periodic differential perturbations, but even pseudo-differential lower-order terms) was made in Parnovski and Sobolev (2010), which also contains a nice review and bibliography of the topic.

  Other related results on BSC can be found in Barbatis and Parnovski (2009); Elton (2004); Lapin (2003); Morozov et al. (2010); Parnovski and Shterenberg (2012); Parnovski and Sidorova (2010); Parnovski and Sobolev (2001a,b); Shen (2003); Sobolev (2007); Vorobets (2011); and in the historical survey in Parnovski and Sobolev (2010).

- **Perturbation Approach: High Frequency Analysis of Dispersion Relation of the Laplacian:** An alternative, **perturbation**, approach to the BSC was developed in Karpeshina (1989, 1990, 1997) and Veliev (1988, 2007, 2024) to prove similar results (see, e.g., Karpeshina, 1997; Veliev, 2024, for detailed references and history). It requires an intricate study of high energy dispersion relation and Bloch eigenfunctions of the Laplacian.

  The average distance between Bloch eigenvalues of the Laplacian (in dimension d) at a high energy λ is about $\lambda^{-(d-2)/2}$. That is, in a multidimensional situation, the eigenvalues are located very densely at high energies. Regular perturbation theory does not work, due to a small denominator problem. However, it can be handled by subtle analytical and geometric considerations (Karpeshina,

1997; Veliev, 2024). It turns out that in the high-energy regime "most" of the eigenvalues and eigenfunctions of the Schrödinger operator with periodic electric potential are close to those of the Laplacian. Let us mention some of the details emerging here:

- There is a "non-resonant" set G of quasi-momenta $\vec{k}$, which is "extensive" in the following sense:

$$|G \cap B_R| = |B_R|(1 + O(R^{-\gamma_1})), \quad \gamma_1 > 0,$$

where B_R is the sphere of radius R centered at the origin, and $|\cdot|$ is the Lebesgue measure in $\mathbb{R}^d$.

The set $\mathbb{R}^d \setminus G$ is called the **resonant set** and plays an important role in crystallography. It has a complicated structure, complexity increasing with dimension. Its simplest part is a vicinity of the so-called **von Laue diffraction planes**, known to physicists for a long time in the case of the cubic lattice with period a:

$$|\vec{k}|^2 = |\vec{k} + 2\pi\vec{q}/a|^2, \quad \vec{q} \in \mathbb{Z}^3.$$

Near these planes the eigenfunctions are not close to plane waves.[14]

- For every $\vec{k} \in G$ there is a generalized eigenfunction $\Psi(\vec{k}, \vec{x})$, which are close to plane waves: for every in an extensive subset G of $\mathbb{R}^d$, $d = 2, 3$, there is a solution $\Psi(\vec{k}, \vec{x})$ of the equation $(-\Delta + V)\Psi = \lambda\Psi$ such that

$$\Psi(\vec{k}, \vec{x}) = e^{i\langle \vec{k}, \vec{x}\rangle} \left(1 + u(\vec{k}, \vec{x})\right), \tag{5.14.2}$$

where $u(\vec{k}, \vec{x})$ is a periodic function and

$$\|u\|_{L^\infty(\mathbb{R}^d)} = O\left(|\vec{k}|^{-\gamma_2}\right), \quad \gamma_2 > 0.$$

- The eigenvalue $\lambda(\vec{k})$ corresponding to $\Psi(\vec{k}, \vec{x})$ is close to $|\vec{k}|^2$:

$$\lambda(\vec{k}) = |\vec{k}|^2 + O\left(|\vec{k}|^{-\gamma_3}\right), \quad \gamma_3 > 0.$$

- Let $D(\lambda)$ be the "nonresonant" part of the Fermi surface of $\lambda(\vec{k})$:

$$D(\lambda) = \left\{\vec{k} \in G : \lambda(\vec{k}) = \lambda\right\}.$$

[14] Physics textbooks describe them as close to linear combinations: $\alpha e^{i\langle \vec{k}, \vec{x}\rangle} + \beta e^{i\langle \vec{k}+\vec{q}/a, \vec{x}\rangle}$. This can be understood as the reflection of a plane wave. It is possible to observe, enabling physicists to find a, and thus define the material. However, theoretical analysis shows that both diffraction conditions and the formula for eigenfunctions are more complicated, see Karpeshina (1995). In particular, the diffraction conditions depend on the potential. In the past, this could not be confirmed in experiments, but with modern equipment, the dependence on the potential can be observed.

Then it is a "slightly distorted" circle with a finite number of holes and can be described by the following formula:

$$D(\lambda) = \left\{ \vec{k} : \vec{k} = \kappa(\lambda, \vec{\nu})\vec{\nu},\ \vec{\nu} \in B(\lambda) \right\},$$

where $B(\lambda)$ is a subset[15] of the unit sphere (circle) S^{d-1}. Here $\kappa(\lambda, \vec{\nu})$ is the "radius" of $D(\lambda)$ in the direction $\vec{\nu}$. The expression $\kappa(\lambda, \vec{\nu}) - \lambda^{1/2}$ describes the deviation of $D(\lambda)$ from the perfect circle of radius $\lambda^{1/2}$. This deviation is small:

$$\kappa(\lambda, \vec{\nu}) = \lambda^{1/2} + O\left(\lambda^{-\gamma_4}\right), \quad \gamma_4 > 0.$$

- The Fermi surface is not empty for every sufficiently large λ, which implies the BSC: the spectrum of H contains a semi-axis.
- The following asymptotics for the DOS $D_V(k^2)$ holds (Karpeshina, 2000):

$$D_V(k^2) = \omega_d k^d + O(k^{-\xi} \ln k) \quad \text{as } k \to \infty,$$

  where $\xi = 3 - d$ and ω_d is a constant depending only on the period lattice of V. The second term of the asymptotics is found for sufficiently smooth potentials for $d = 3$.
- It has been noted (see, e.g., Feldman et al., 1990, 1991; Friedlander, 1990, 1991) that the Floquet eigenvalues $\lambda_j(k)$ can be split into the sets of "nice" **stable** and more complicated **unstable** or **resonant** ones. The reader is directed to the works cited for technical details and further references.

• **A stronger version of BSC:** This applies to any second-order periodic elliptic operator. It is still not proven, except for operators of the form

$$H = (-\Delta)^m + \text{ lower order periodic terms;} \tag{5.14.3}$$

see Parnovski (2008) and Parnovski and Sobolev (2010), and references therein. Major difficulties appear to lie in handling variable coefficients in the highest order terms. The author is unaware of any progress in this direction.

One can find a **semi-classical approach to the BSC** in Ivrii (2019a).

• **Periodic Laplace–Beltrami in higher dimension:** As we have seen already in the 1D case, properties of a **periodic Laplace–Beltrami operators** can differ significantly from those of Schrödinger ones. For example, in 1D the former ones do not have any spectral gaps, while the latter ones normally have infinitely many of them. In higher dimensions, the situation in some sense reverses. Namely, 2D examples of periodic Laplace–Beltrami operators with arbitrarily many spectral gaps were constructed in Green (1997). The question on whether infinitely many

[15] The set $B(\lambda)$ can be interpreted as the set of an asymptotically full (as $\lambda \to \infty$) measure on S^{d-1} of possible directions of propagation for almost plane waves (5.14.2).

gaps are possible (i.e., about the validity of BSC) for such operators seems to be still open.

It is interesting to note that the method used in Green (1997) to show that the number of possible gaps in spectra of periodic Laplace–Beltrami operators in 2D can be arbitrarily large is essentially the same one that was applied in Figotin and Kuchment (1996a,b, 1998) for showing existence of gaps in the spectra of some 2D photonic crystals (see Kuchment, 2001, 2027).

- **BSC for other operators:**
  - The validity of **BSC for periodic Maxwell operators** is mostly unresolved, except for a very special case considered in Vorobets (2011).
  - The validity of **BSC for periodic waveguide systems** has not been established and is still in question.
  - For **periodic graphs/quantum graphs** the BSC does not hold (Berkolaiko and Kuchment, 2013; Kuchment, 2016).
- **Ballistic versus spectral properties:**

  Deviating from the main topic of this book, we briefly mention the following topic, which will belong to Kuchment (2027). Namely, the spectral and dynamical properties of Schrödinger operators $H = -\Delta + V$ are related. A general correspondence of this kind is given by the RAGE theorem (Simon, 1979) (see also the survey of Damanik, 2017). Stated briefly, it says that solutions $\Psi(\cdot,t) = e^{-iHt}\Psi_0$ of the time-dependent Schrödinger equation are **bound states** if the spectral measure μ_{Ψ_0} of the initial state Ψ_0 is pure-point and a **scattering state** if μ_{Ψ_0} is (absolutely) continuous.

  More precise quantification of the transport properties is done in terms of **diffusion exponents** β, which characterize how time-averaged m-moments ($m > 0$) of the position operator X

$$\langle\langle X^m_{\Psi_0}\rangle\rangle_T := \frac{2}{T}\int_0^\infty \exp\left(-\frac{2t}{T}\right) \|X^{m/2}\|\Psi(\cdot,t)^2_H \, dt$$

  grow as a power $T^{m\beta}$ of time T, where $(Xu)(x) = |x|u(x)$. The special cases $\beta = 1$, $\beta = 1/2$, and $\beta = 0$ reflect respectively the **ballistic transport, diffusive transport, and dynamical localization**.

  The ballistic upper and lower bounds for periodic (and sometimes more general) potentials can be found in Asch and Knauf (1998); Kiselev and Last (2000); and Radin and Simon (1978).
- **Section 5.5** The role of **spectral decomposition into generalized eigenfunctions**: See, for example, Berezanskii (1968); Berezin and Shubin (1991); and Gel′fand and Kostyučenko (1955). Although not discussed much in this text,

has relations to various topics that we have considered, for example, the limiting absorption principle, the Shnol' theorem, and others.

- **Section 5.7 Absolute continuity:** As we have discussed, the issue of absence of bound states (i.e., absolute continuity of the spectrum) would have been resolved if the so-called **Landis conjecture** were proven. See more details about this in Appendix E.

  Absolute continuity of the spectrum of isotropic periodic Maxwell operator was proven in Morame (2000) by a nice reduction to a **matrix** Schrödinger operator. In Kuchment and Levendorskiî (2002), there was an attempt made to extend this result to the matrix $\mathbb{Z}^d$-periodic Schrödinger operators in $\mathbb{R}^d$ acting on n-vector functions incorporating matrix "magnetic potentials":

$$H = -\begin{pmatrix} \Delta & 0 & \cdots & 0 \\ 0 & \Delta & 0 & \cdots \\ \cdots & \cdots & \cdots & \cdots \\ 0 & 0 & \cdots & \Delta \end{pmatrix} - 2i\sum_{j=1}^{d} A_j(x)\frac{\partial}{\partial x_j} + V(x), \qquad (5.14.4)$$

  where the matrix "magnetic and electric potentials" $A_j(x)$ and $V(x)$ are periodic with some smoothness conditions imposed upon A_j. **Absence of L_2-eigenfunctions** is proven under some additional algebraic conditions on the potentials. In the self-adjoint case (notice that self-adjointness was not required), this implies **absolute continuity of the spectrum**.

  Here are the algebraic conditions that we have mentioned. Besides requiring the magnetic potentials to average to zero, the following commutation relation was required:

  Condition *For any $j, k = 1, \ldots, d$ and $x, y \in \mathbb{R}^d$, the following commutation relations needs to be satisfied:*

$$[A_j(x), A_k(y)] + [A_k(x), A_j(y)] = 0. \qquad (5.14.5)$$

  At that moment, all cases (with constant coefficients in the leading term) known to the author, including the result of Morame (2000), reduced to this form. However, it is still not clear whether the relation (5.14.5) is indeed needed to avoid counter-examples, or it was just an artifact of the proof.

- **Section 5.8**: The original statement (and thus the proof) of **Shnol' theorem**, requiring exponential estimate

$$|u(x)| \leq Ce^{a|x|} \qquad (5.14.6)$$

  on the generalized eigenfunction u is of no use for some other manifolds, for example, on those with super-polynomial growth of the ball volume, such as hyperbolic manifolds or regular trees. However, the issue turns out to be not with much the proof, but rather with too restrictive statement. The widely applicable estimate on generalized eigenfunction should be as follows: if B_r is a ball of radius r, then the condition should be

$$\|u(x)\|_{L_2(B_r)} \leq Ce^{ar}.$$

With this formulation, the classical proof stays essentially the same, but the result now works without any conditions on the volume growth. One can find a more detailed discussion in the quantum graph case in Berkolaiko and Kuchment (2013).

- **Section 5.9:** In formula (5.9.5), one encounters the **integral of a holomorphic form over a real cycle** on the Fermi surface F_λ, which allows one to use powerful techniques of several complex variables, which we are unable to address here; see Gérard (1990); Gieseker et al. (1993, 1991); Pham (1967).
- **Section 5.10:** Concerning the role of **Van Hove singularities** in the spectral theory, see, for instance, Colin de Verdière (1991); Gérard (1990).
- **Section 5.11:** One can find interesting **S. Novikov's conjecture on the semi-classical electron** and its resolution in Dynnikov (1992, 1993b,a, 1996, 1997, 1999); Novikov (1999).
- **Section 5.13:** Concerning **inverse problems for periodic operators**, one should also see Eskin (1989); Eskin et al. (1984b,a,c); Gordon and Kappeler (1991, 1995); Gordon et al. (2008); Guillemin (1990); and Veliev (2009, 2011).

6

Localized Perturbations of Periodic Operators

6.1 Introduction

Consider the stationary Schrödinger operator

$$H_0 = -\Delta + V(x) \tag{6.1.1}$$

in $L^2(\mathbb{R}^n)$ with the real potential $W(x) \in L^\infty(\mathbb{R}^n)$ periodic with respect to the integer lattice $\mathbb{Z}^n$:

$$V(x + l) = V(x), l \in \mathbb{Z}^n, x \in \mathbb{R}^n.$$

As we know from Sections 4.6.1, 5.2, and 5.7, the spectrum of this operator is absolutely continuous and has a band-gap structure:

$$\sigma(H_0) = \bigcup_{i \geq 1} [a_i, b_i],$$

where $a_i < b_i$, and $\lim_{i \to \infty} a_i = \infty$. Let us introduce also a perturbed operator with a **perturbation** (or **impurity**) being a sufficiently localized potential[1] $v(x)$:

$$H = H_0 + W(x) = -\Delta + V(x) + W(x), \tag{6.1.2}$$

where the function $W(x)$ is compactly supported or sufficiently fast decaying at infinity (the exact conditions can vary and will be discussed later). According to Weyl's theorem (Weyl, 1909) (see also Theorem 5.35 in Kato, 1995), if the perturbation decays sufficiently fast so as to be relatively compact, the essential spectrum of H coincides with the spectrum of H_0, and only some additional "impurity" point spectrum $\{\lambda_j\}$ can arise.[2]

The results for the 1D case are mostly summarized in Theorem 1.6.1. There are more possibilities in the higher dimension. The newly arising impurity spectrum in

[1] Such a perturbation might indicate a local deviation from a perfect periodic structure, for example, a "foreign" atom.

[2] See also Brown et al. (2013); Glazman (1966, Section 18); Reed and Simon (1972); and references therein.

a gap can be finite or infinite and can appear both in the gaps (including their edges) and in the interior of the essential spectrum of the operator (in the latter case, one talks about the **embedded eigenvalues**). There is extensive technical literature on this topic, so addressing even the major details and techniques would have required a large survey in itself or even a separate book. Thus, the author has chosen to provide just a brief overview of the history and the current state of the topic with references to the relevant sources.

6.2 Perturbations

As we have agreed, we will consider "sufficiently fast" decaying perturbations of a periodic potential. The meaning of "sufficiently fast" decay might be different, which can (and does) influence the results.

We will confine ourselves to the case of the Schrödinger operator, albeit many results can be translated to more general second-order periodic elliptic operators. So, let us consider the unperturbed operator

$$H_0 := -\Delta + V(x) \tag{6.2.1}$$

with (Γ-)periodic potential $V \in L_\infty$, which is assumed for simplicity to be bounded, and its perturbation

$$H := -\Delta + V(x) + W(x) = H_0 + W(x). \tag{6.2.2}$$

We assume that both V and W are real-valued and (with some overkill) that the perturbation $W(x)$ is continuous and $\lim\limits_{\|x\|\to\infty} W(x) = 0$.

Theorem 6.2.1 *Under these conditions, the Schrödinger operator H is self-adjoint and $\sigma_{\text{ess}}(H) = \sigma_{\text{ess}}(H_0)$.*

6.3 Impurity Spectra

Let us return now to the periodic Schrödinger operator $H = -\Delta + V(x)$. An important and frequently studied issue is the existence and properties of the **impurity spectrum** arising when a localized (compactly supported or sufficiently fast decaying) perturbation $W(x)$ is added to the potential. Weyl's theorem mentioned earlier claims that only eigenvalues of finite multiplicity can appear, leaving the otherwise absolutely continuous spectrum unchanged (see, e.g., Section 18 of Glazman, 1966 and references therein).

According to Theorem 6.2.1, a relatively compact perturbation may (and often does) lead to appearance of "impurity" discrete spectrum. Since V is periodic and thus has band-gap structure of the spectrum, these "impurity" eigenvalues have two options: to appear in the spectral gaps or embed into the AC spectrum (**embedded eigenvalues**) (see Fig. 6.1).

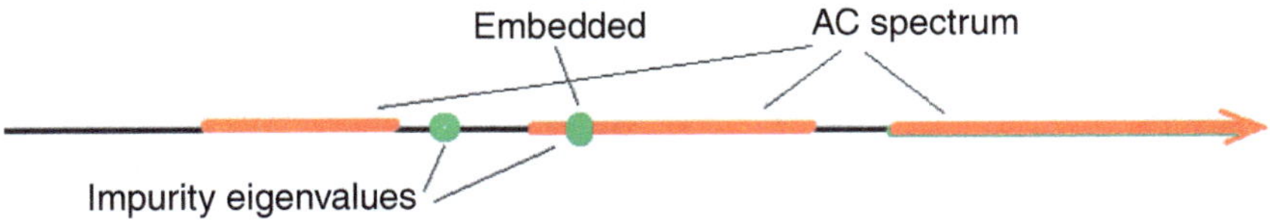

Figure 6.1 Impurity eigenvalues on a band-gap structured spectrum.

We will address both cases in the following subsections.

6.3.1 Spectrum in Gaps

Active study of the impurity eigenvalues in the spectral gaps was largely triggered by the classical works of Alama et al. (1989) and Hempel (1989, 1992) (see the overview in Hempel, 2008). In particular, a lot of attention has been devoted to studying the case of a periodic elliptic operator perturbed by a decaying potential, a topic that fits well into this text.

From now on, we will deal with the unperturbed and perturbed operators H_0 and H as in Section 6.2.

Let (α, β) be a finite or infinite gap in the spectrum; in the latter case, $\alpha = -\infty$ and the gap is semi-infinite; otherwise the gap is finite. The case of the semi-infinite gap has been studied a lot (see, for instance, the introduction and references in Birman et al., 2000b). We thus assume here that the gap is finite. The case of nonnegative (or nonpositive) perturbation $W(x)$ is better understood, so we assume that $W(x) \geq 0$. Then the possible eigenvalues of the operator $H_0 + tW$ can emerge from α and move up through the gap when t increases. We thus can define for $\lambda \in (\alpha, \beta)$ the **counting function**:

$$N(\lambda, \tau, W) := \#\{ \text{ eigenvalues crossing } \lambda \text{ for } t \in (0, \tau)\}. \tag{6.3.1}$$

We consider now a more general operator in $\mathbb{R}^d$:

$$H_0 := -\nabla \cdot g(x)\nabla + V(x), \tag{6.3.2}$$

where $V \in L_\infty(\mathbb{R}^2)$ and $g(x)$ is a real-valued $d \times d$-matrix function with entries in $L_\infty(\mathbb{R}^d)$ such that $g(x) \geq c\mathbf{I}$ with $c > 0$. Both g and V are assumed periodic w.r.t. the integer lattice.

Let us collect the **assumptions** we will need:

$$d \geq 1. \tag{6.3.3}$$

$$\alpha \text{ is an isolated maximum value of the band function } \lambda_{l-1}. \tag{6.3.4}$$

$$\beta \text{ is a minimum value of the band function } \lambda_l. \tag{6.3.5}$$

$$\lambda \in (\alpha, \beta) \tag{6.3.6}$$

$$W(x) \geq 0, W \in L_\infty(\mathbb{R}^d) \tag{6.3.7}$$

$$W(x) = \omega(x/|x|)|x|^{-\sigma} + o(|x|^{-\sigma}), |x| \to \infty \text{ for some } \sigma \in (0, d). \tag{6.3.8}$$

We will also need the following.

Definition 6.3.1

$$\Gamma_\sigma(\lambda) := \frac{1}{d(2\pi)^d} \sum_{s=1}^{l-1} \int_{\mathcal{B}} (\lambda - \lambda_s(k))^{-d/\sigma}\, dk \int_{S^{d-1}} \omega^{d/\sigma}(\theta)\, dS(\theta). \tag{6.3.9}$$

Now we can formulate the result stated in Birman (1996) and proven in Birman and Sloushch (2010, Theorem 1.3).

Theorem 6.3.2 *Under assumptions* (6.3.3)–(6.3.8)*, the following asymptotics holds:*

$$\lim_{\tau \to \infty} \tau^{-d/\sigma} N(\lambda, \tau, W) = \Gamma_\sigma(\lambda). \tag{6.3.10}$$

Remark 6.3.3

- For $g = \mathbb{I}$, this result was obtained in the pioneering work of Alama et al. (1989), with periodicity of V replaced by the condition of existence of integrated density of states for $-\Delta + V$.
- One wonders what happens at the gap edge α (which thus is a [local] maximum of the dispersion relation). This happens to be a much trickier question, especially in dimension $d = 2$, and the answer depends on whether the perturbation is "regular" or "irregular." In the latter case, the structure of the dispersion relations near its maximum α starts playing a role (see Birman, 1995, 1996; Birman and Sloushch, 2010; Raĭkov, 1992 for the details). The 2D case happens to be special and more difficult to study (Birman et al., 2000b; Suslina, 2003b).
- If the perturbation changes the sign, there are apparently no exact asymptotics known for the impurity eigenvalues in a gap, but rather some estimates (see, e.g., Levendorskiĭ, 1995b).
- A nice approach to decaying perturbations of periodic operators is to consider them (after the Floquet transform) as pseudo-differential operators with operator symbols (Levendorskiĭ, 1995a).

- The case of matrix perturbation $W(x)$ was considered in Ivrii (1999). Perturbations by differential operators have also been considered (see, e.g., Birman, 1994; Boyarchenko and Levendorskiĭ, 1997a).
- The reader interested in this topic should also look at Birman and Weidl (1994); Boyarchenko and Levendorskiĭ (1997a,b); Hoefer and Weinstein (2011); Kamotski and Smyshlyaev (2018); Levendorskiĭ (1985, 1996, 1997).

6.3.2 Embedded Eigenvalues

Concerning eigenvalues possibly appearing in the interior of a spectral band of $\sigma(H_0)$ (which we will call **embedded** eigenvalues), there are known examples of such eigenvalues that arise when the perturbation does not decay sufficiently fast, first of which was found in 1D by J. von Neumann and E. Wigner (1929); see Eastham and Kalf (1982) and Reed and Simon (1972, Section XIII.13) for discussions and related references. However, a general expectation is that they are unlikely to arise, as long as the perturbation decays sufficiently fast (e.g., is compactly supported). This issue has been intensively studied. There are many known results on nonexistence of embedded eigenvalues for the case of absence of the underlying periodic potential $V(x)$ (see, for instance, Eastham and Kalf, 1982, Section 14.7 of Hörmander, 2005, and Section XIII.13 of Reed and Simon, 1972). Apparently, until the last couple of decades, the only known results on the absence of embedded eigenvalues in the periodic case have been established in 1D (see Chapter 1; Rofe-Beketov, 1977, 1984, 1973, 1964b; and Section 6.4).

So, our goal is to obtain a result of this kind for higher dimensions. It is generally believed that in any dimension, if the perturbation decays fast enough (e.g., exponentially), no embedded eigenvalues should arise. This, however, happens to be a hard nut to crack in periodic case, due to its relation (shown in the following) to intricate analytic properties of the Fermi surface.

Now we introduce our basic condition:

Condition 6.3.4 *Assume that for a given λ that belongs to the interior of a spectral band of the Schrödinger operator,*

$$H_0 = -\Delta + W(x),$$

every irreducible component of the Fermi variety $F_\lambda(q)$ intersects the real space $\mathbb{R}^n$ along a subset of dimension $(n-1)$ (i.e., a subset that contains a piece of a smooth hypersurface).

Theorem 4.6.11 says that for λ in the interior of a spectral band, the Fermi variety $F_\lambda(q)$ does intersect the real space $\mathbb{R}^n$ over a subset of dimension $(n-1)$. We, however, need more, that **every irreducible component of $F_\lambda(q)$ does the same.**

In other words, there are no "hidden" components that do not show up on the real subspace in any significant way. Thus, Theorem 4.6.11 and irreducibility of $F_\lambda(q)/2\pi\mathbb{Z}^n$ would imply validity of Condition 6.3.4.

We can now prove our main (conditional) statement.

Theorem 6.3.5 *If a real periodic potential $V(x) \in L^\infty(\mathbb{R}^n)$ $(n \leq 3)$ satisfies Condition 6.3.4 for some λ and an impurity potential $W(x)$ is measurable and satisfies the estimate,*

$$|W(x)| \leq Ce^{-|x|^r} \text{ for some } r > 4/3 \tag{6.3.11}$$

almost everywhere in $\mathbb{R}^n$, then λ cannot be an embedded eigenvalue in the spectrum of H.

Before providing the proof, we mention an immediate corollary:

Corollary 6.3.6 *If a real periodic potential $V(x) \in L^\infty(\mathbb{R}^n)$ $(n \leq 3)$ satisfies the Condition 6.3.4 for **all** λ and an impurity potential $W(x)$ is measurable and satisfies the estimate,*

$$|W(x)| \leq Ce^{-|x|^r} \text{ for some } r > 4/3 \tag{6.3.12}$$

almost everywhere in $\mathbb{R}^n$, then the spectrum of H contains no embedded eigenvalues. In other words,

$$\{\lambda_j\} \cap \left(\bigcup_{i \geq 1} (a_i, b_i) \right) = \emptyset,$$

where $\{\lambda_j\}$ is the impurity point spectrum of H, and

$$\bigcup_{i \geq 1} [a_i, b_i] = \sigma(H_0)$$

is the band structure of the essential spectrum of H.

Proof of Theorem 6.3.5 Let us assume that there exists a value λ of the spectral parameter that belongs to some (a_i, b_i) and to the point spectrum of H simultaneously. Then there exists a nonzero function $u(x) \in L^2(\mathbb{R}^n)$ (an eigenfunction) such that

$$-\Delta u + Vu + Wu = \lambda u,$$

or

$$(H_0 - \lambda)u = -Wu.$$

Let us denote the function in the right-hand side by $\psi(x)$:

$$\psi(x) = -W(x)u(x).$$

Consider the fundamental domain $\mathbf{W}$ of the group $\mathbb{Z}^n$ of periods:

$$\mathbf{W} = \{(x_1, \dots, x_n) \in R^n \mid 0 \le x_i \le 1,\ i = 1, \dots, n\}.$$

Then (6.3.12) implies that the function $\psi(x)$ satisfies the estimate,

$$\|\psi\|_{L^2(\mathbf{W}+l)} \le Ce^{-|l|^r} \tag{6.3.13}$$

for all $l \in \mathbb{Z}^n$, where $r > 4/3$. Consider the Floquet transform of function $\psi(x)$. We get a function of $k \in \mathbb{C}^n$

$$\phi(k) = \widehat{\psi}(k, x) \tag{6.3.14}$$

with values in the space $L^2(\mathbb{T})$.

Lemma 6.3.7 *The function $\phi(k)$ is an entire function on $\mathbb{C}^n$ of the order $s = r/(r-1) < 4$.*

Proof of the Lemma From (6.3.13), we get

$$\left\|\psi(x-l)e^{-ik\cdot(x-l)}\right\|_{L^2(\mathcal{K})} \le Ce^{C|k|+\mathrm{Im}(k\cdot l)-|l|^r}.$$

Then

$$\|\phi(k)\|_{L^2(\mathcal{K})} \le Ce^{C|k|} \sum_{l\in\mathbb{Z}^n} e^{-0.5|l|^r} e^{\mathrm{Im}(k\cdot l)-0.5|l|^r}.$$

A simple Legendre transform type estimate (finding extremal values of the exponent) shows that

$$e^{\mathrm{Im}(k\cdot l)-0.5|l|^r} \le Ce^{C|k|^{r/(r-1)}},$$

which proves that the function $\phi(k)$ is an entire function of the order $s = r/(r-1)$ in $\mathbb{C}^n$. □

As we know from Section 3.6, the Floquet transform $\mathcal{U}_\Gamma$ leads to the following operator equation with a parameter k:

$$(H_0(k) - \lambda)\,\widehat{u}(k) = \phi(k), \tag{6.3.15}$$

where

$$\widehat{u}(k) = (\mathcal{U}_\Gamma u)\,(k, x)$$

is defined for $k \in \mathbb{R}^n$ and

$$H_0(k) = (i\nabla - k)^2 + W(x)$$

is considered as an operator on the torus $\mathbb{T}$. Here

$$\widehat{u}(k) \in L^2_{\mathrm{loc}}(\mathbb{R}^n, L^2(\mathbb{T})).$$

In fact, standard interior elliptic estimates show that

$$\widehat{u}(k) \in L^2_{\text{loc}}(\mathbb{R}^n, H^2(\mathbb{T})),$$

where $H^2(\mathbb{T})$ is the standard Sobolev space of order two on $\mathbb{T}$. The operator

$$(H_0(k) - \lambda) : H^2(\mathbb{T}) \to L^2(\mathbb{T}^n)$$

is invertible if and only if $k \notin F_{\lambda,\mathbb{R}}(q)$. As is shown in Section B.5, the inverse operator can be represented as the ratio

$$(H_0(k) - \lambda)^{-1} = B(k)/\zeta(k), \tag{6.3.16}$$

where $B(k)$ is a bounded operator from $L^2(\mathbb{T})$ into $H^2(\mathbb{T})$, and $B(k)$ and $\zeta(k)$ are correspondingly an operator and a scalar entire functions of order n in $\mathbb{C}^n$. Furthermore, the zeros of $\zeta(k)$ constitute exactly the Fermi variety $F_\lambda(q)$. We conclude now that the following representation holds on the set $\mathbb{R}^n \backslash F_{\lambda,\mathbb{R}}(q)$:

$$\widehat{u}(k) = \frac{B(k)\phi(k)}{\zeta(k)} = \frac{g(k)}{\zeta(k)},$$

where $g(k)$ is a $H^2(\mathbb{T}^n)$-valued entire function of order n.

Now we need the following auxiliary result:

Lemma 6.3.8 *Let Z be the set of all zeros of an entire function $\zeta(k)$ in $\mathbb{C}^n$ and Z_j be its irreducible components. Assume that the real part $Z_{j,\mathbb{R}} = Z_j \cap \mathbb{R}^n$ of each Z_j contains a submanifold of real dimension $(n-1)$. Let also $g(k)$ be an entire function in $\mathbb{C}^n$ with values in a Hilbert space H such that on the real subspace $\mathbb{R}^n$ the ratio*

$$\widehat{u}(k) = \frac{g(k)}{\zeta(k)}$$

belongs to $L_{2,\text{loc}}(\mathbb{R}^n, H)$. Then $\widehat{u}(k)$ extends to an entire function with values in H.

Proof In the first step, one shows that the ratio g/ζ is analytic at the points of the real space. In other words, zeros of g must cancel zeros of ζ there. Indeed, otherwise one would deal with a non-square-integrable singularity. Then, in the second step, since all irreducible components of zeros of ζ have significant (of dimension $n-1$) presence on the real space, an analytic continuation argument shows that the cancellation works in the complex domain as well. □

Using now the result of Theorem 4.6.11 and the assumption that Condition 6.3.4 is satisfied for the potential $W(x)$, we conclude that $\widehat{u}(k)$ is an entire function and that it is the ratio of two entire functions of order at most $w = \max(n, s) < 4$, where s is defined in Lemma 6.3.7. Using (in the radial directions) the estimate of entire functions from below contained in Levin (1996, chapter 1, section 8) – see also

Kuchment (1993, Theorem 1.5.6 and Corollary 1.5.7) or similar results in Boas (1954) – we conclude that $\widehat{u}(k)$ is itself an entire function of order w with values in $H^2(\mathbb{T})$. Now the Paley–Wiener type theorem (Theorem 3.4.20) claims that the solution $u(x)$ satisfies the decay estimate:

$$||u||_{H^2(K+a)} \le C_p \exp(-c_p \, |a|^p)$$

for any $p < w/(w-1)$, where K is an arbitrary compact in $\mathbb{R}^n$, and $a \in \mathbb{R}^n$. Using standard embedding theorems, we conclude that

$$|u(x)| \le C_p \exp(-c_p \, |x|^p) \text{ for any } p < w/(w-1).$$

Since $w < 4$, one can choose a value p such that

$$4/3 < p < w/(w-1).$$

However, the results concerning possible rate of decay of solutions that are contained in Froese et al. (1983, Remark 2.6) and Meshkov (1991, Theorem 1) state that there is no nontrivial solution of the equation

$$-\Delta u + Vu + Wu - \lambda u = 0$$

with the rate of decay

$$|u(x)| \le C \exp(-c \, |x|^p), p > 4/3.$$

This contradiction concludes the proof of the theorem. □

This result leads to the problem of finding classes of periodic potentials that satisfy Condition 6.3.4. As we stated in Conjectures 4.6.34 and 4.6.35, we believe that all (or almost all) periodic potentials satisfy this condition. Although we were not able to prove these conjectures, as an immediate corollary of Theorems 4.6.36 and 6.3.5, we get the following result:

Theorem 6.3.9 *If, for $n < 4$, the background $\mathbb{Z}^n$-periodic potential $W(x) \in L^\infty(\mathbb{R}^n)$ is separable for $n = 2$ or $W(x) = q_1(x_1) + q_2(x_2, x_3)$ for $n = 3$, and the perturbation potential $v(x)$ satisfies the estimate* (6.3.12)*, then there are no eigenvalues of the operator H in the interior of the bands of the continuous spectrum.*

The reader should notice that the separability required from the potential goes along the axes of the lattice of the periods. This condition is most surely irrelevant.

As we have mentioned before, the general wisdom seems to be that if the perturbation decays sufficiently fast, embedded eigenvalues cannot arise, while if the decay is not fast enough, one can indeed have them. Many results of this nature have been obtained in 1D, see, for example, Rofe-Beketov (1963, 1964b, 1973, 1984) and a nice survey in Eastham and Kalf (1982).

Although similar answers are expected in the higher dimensions, proving them happens to be difficult. Some results were obtained in Kuchment and Vainberg (1998, 2000, 2006). For example,

Theorem 6.3.10 (Kuchment and Vainberg, 1998, 2000) *Let $n \leq 4$ and the perturbation $W(x)$ decays exponentially. If for a given λ in the interior of $\sigma_{\text{ac}}(H)$, the Fermi surface at that level is irreducible (modulo periodicity), then $\lambda \notin \sigma_{\text{pp}}(H+W)$. In particular, if all Fermi surfaces are irreducible, no embedded eigenvalues can arise.*

Remark 6.3.11 Notice that irreducibility of the Fermi surface at a level does not depend on the impurity potential W.

As our previous discussion has indicated, it is hard to establish irreducibility of the Fermi surface, although for a second-order periodic ODE, it is automatic. One wonders whether the irreducibility condition of the theorem mentioned earlier is just an artifact of the techniques used in the proof. This is possible but does not seem to be very likely. The proof indicated that reducibility could in principle lead to a situation similar to the waveguide theory, where impurity eigenvalues can get embedded. And indeed, there are examples of a periodic fourth-order ODEs (Papanicolaou, 2021) and of second-order multi-periodic quantum graph operators (Shipman, 2014), where reducible Fermi surfaces arise and impurity eigenvalues can be embedded. The idea is to create an embedded impurity eigenfunctions using only the "hidden" in complex domain component of the Fermi surface.

In the graph case (which we do not address in this text), due to nontrivial topology, embedded eigenvalues can and do arise.[3] However,

Theorem 6.3.12 (Kuchment and Vainberg (2006)) *In the graph case, embedded eigenvalues might arise, but if the relevant Fermi surface is irreducible, the corresponding eigenfunction must be supported "close" to the support of the perturbation.*

6.4 Comments and References

- **Section 6.3.1:**
  - The classical Weyl's theorem (Weyl, 1909) can be found in essentially any book on spectral theory, for example, Reed and Simon (1972, Vol. 4) or Kato (1995).
  - As we have mentioned, many papers have been devoted to studying the behavior of the point spectrum in the gaps of the continuous spectrum; see, for example, Alama et al. (1994, 1989); Birman (1991a,b, 1994, 1995, 1996,

[3] It is interesting to note that in such case, eigenfunctions can spread and indeed have unbounded support (Shipman, 2014).

1997); Birman and Raĭkov (1991); Birman and Sloushch (2010); Birman and Weidl (1994); Birman et al. (2000b); Boyarchenko and Levendorskiĭ (1997a,b); Deift and Hempel (1986); Gesztesy and Simon (1988); Hempel (1989, 1992, 2008); Hempel and Laitenberger (1994); Hempel and Levendorskiĭ (1998); Levendorskiĭ (1995b, 1997); Raikov (1998); Rofe-Beketov (1964a,b); Sobolev (1991); Suslina (2003b,a).

- **Section 6.3.2:**
  - In our considerations, we follow an approach that was used in Vaĭnberg (1966) for the case of zero underlying potential $V(x)$.
  - In Theorem 6.3.10, we only prove the absence of eigenvalues embedded into the interior of a spectral band. It is possible that eigenvalues cannot occur at the ends of the bands either, if the perturbation potential decays fast enough. This was shown in the 1D case under the condition

$$\int (1 + |x|)\, |v(x)|\, dx < \infty$$

    on the perturbation potential. If $W(x)$ only belongs to L^1, the eigenvalues at the endpoints of spectral bands can occur (see more details and references in Chapter 1). The author is unaware of such results in higher dimensions.
  - Most of the proof of the conditional Theorem 6.3.5 applies to periodic elliptic operators more general than the Schrödinger operator with an electric potential. The only obstacle (albeit major) occurs at the last step, when one needs to conclude the absence of too fast (super-exponentially) decaying solutions to the equation (**Landis conjecture**, see Appendix E). Any progress in this direction would automatically generalize Theorem 6.3.5.
  - The restriction on the spatial dimension $n < 4$ comes from the allowed rate of decay stated in the result of Froese et al. (1983); Meshkov (1991) (the notorious $4/3$). There is an example provided in Meshkov (1991) showing that this constant is sharp. However, the example involves complex-valued potentials. There is a point of view shared by some researchers that in the self-adjoint case, no super-exponential decay whatsoever should be allowed, and thus 1 would replace $4/3$. If this were true, our dimension restriction would be removed.[4]
  - It might seem that the irreducibility condition 6.3.4 is an artifact of the proof and is only sufficient, but may not be necessary. We believe this is not to be the case, and that the irreducibility is essentially "equivalent" to the absence of embedded eigenvalues, as the examples provided in the works mentioned previously seem to confirm.

[4] A warning sign comes from the result of Krymskii and Filonov (2023), where $4/3$ is attained in a self-adjoint case but in a **half**-cylinder.

7

Wannier Functions

7.1 Introduction: A Missing Basis

As the reader who has reached this point must be already persuaded, Floquet–Bloch functions and Floquet transform play in the periodic case the roles of exponential plane waves and Fourier transform correspondingly. We recall that Bloch functions have the form $\phi(k,x) := e^{ik\cdot x}p(x)$ with periodic $p(x)$. Here the quasi-momentum k plays the role of the dual variable ξ.[1]

When dealing with strongly localized functions, the Fourier basis involves expansions with very long tails. This is the case, for instance, when one deals with "almost discrete" systems, such as a crystal, which can be viewed as a large "empty" space sparsely populated by tiny atoms. In such cases, a basis of strongly localized functions might seem to be more helpful.

Let us take look at the homogeneous medium case (arising in the situations when dealing with constant coefficient linear differential operators). We recall that two extremely useful function bases exist. Namely the delta-functions $\delta_a(x) := \delta(x-a)$ (localized in the space) and exponentials (plane waves) $e_\xi(x) := e^{ix\cdot\xi}$ (localized in the dual variable, momentum ξ). They are related by means of Fourier transform. Indeed, the standard Fourier transform maps the first set into the second one, and similarly the first one is obtained by inverse Fourier transform of the second ones.

We have already found in periodic case analogs of the plane waves (Bloch functions) and of the Fourier transform (Floquet transform (3.2.4)), which transforms functions of x into functions of k with values in functions of x:

$$f(x) \mapsto \mathcal{U}_\Gamma f(x,k) := \widehat{f}(x,k) := \sum_{\gamma\in\Gamma} f_\gamma(x)e^{ik\cdot\gamma} = \sum_{\gamma\in\Gamma} f_\gamma(x)z^\gamma, \qquad (7.1.2)$$

[1] Notice that the periodic factor $p(x)$ may (and usually will) also depend on k:

$$\phi(k,x) = e^{ik\cdot x}p(k,x). \qquad (7.1.1)$$

where, as before $z = e^{ik}$. Alas, we seemingly do not have yet for the periodic case analogs of bases localized in space, while they could be extremely useful for studying crystal structures, where basic functions significantly localized near atoms would be very useful for numerical analysis.[2]

It looks, however, that we have all tools needed to introduce the missing analog of the spatially localized (in the best possible way, preferably exponentially[3]) function system. Namely, analogy with the homogeneous case suggest to apply the inverse Floquet transform to Bloch functions! This is exactly what is attempted in this chapter, introducing as the result the **Wannier functions** and studying nontrivial questions of their existence, decay, and basis property.

The resulting fast decaying Wannier functions are used frequently in numerical studies in solid-state physics and photonic crystal theory (Busch et al., 2003; Kohn, 1959, 1973; Rehr and Kohn, 1974; Schillinger et al., 2005; Thouless, 1984; Wannier, 1937), since they lead to sparse systems, akin to the tight-binding models. In particular, a major and extremely influential variational approach numerical program for constructing "Optimal Wannier functions" has been created (Brouder et al., 2007; Damle et al., 2018; Marzari and Vanderbilt, 1997; Marzari et al., 2003, 2012; Mostofi et al., 2008, 2014; Pizzi et al., 2020; Soluyanov and Vanderbilt, 2011; Souza et al., 2001; Yates et al., 2007). Analytic understanding of why, when, and how well this approach works is still being developed.

7.2 What Are Wannier Functions?

Thus, in analogy with the Euclidean Fourier analysis, we consider inverse Floquet transforms of Bloch functions. Using the inversion formula (3.2.10), one comes to the following notion:

Definition 7.2.1 Let $\phi(k,x)$ be a Bloch function (see (7.1.1)) that depends "sufficiently nicely" on the quasi-momentum k. The corresponding **Wannier function**[4] is defined as the inverse Floquet transform of ϕ:

$$w(x) := \int_{\mathbb{T}^*} \phi(k,x)d'k, \tag{7.2.1}$$

2 "True" localization, that is, having compact support is impossible for solutions of second order elliptic equations, due to the weak uniqueness of continuation property mentioned before.

3 We repeat again that the weak uniqueness of continuation property prohibits us to require compactness of support of solutions.

4 In the physics literature, the name "Wannier function" is usually reserved for functions coming from a Bloch **eigenfunction** ϕ, which we do not assume. In our more general situation, the terms **generalized (or composite) Wannier function** is used. We will use the shorter term Wannier function in this extended sense.

and correspondingly $\phi(k,x)$ is the Floquet transform of $w_(x)$:

$$\phi(k,x) = \widehat{w}(k,x) = \sum_{\gamma\in\Gamma} w(x+\gamma)e^{-ik\cdot\gamma}. \tag{7.2.2}$$

Our goal is to find sufficiently fast decaying functions w. How do we make sure that the such defined Wannier function decays (and how fast can it decay)? Here the previously established Paley–Wiener type theorems for the Floquet transform describing the decay of $w(x)$ in terms of smoothness[5] of $\phi(k,\cdot)$ with respect to k come handy. This is done, in particular, in Theorem 3.4.5, which we repeat here for convenience:

Theorem 7.2.2 (Repeat of Theorem 3.4.5) *For any $a \in (0,\infty]$, the Floquet transform*

$$\mathcal{U}_\Gamma : f \mapsto \hat{f}(z,x)$$

is a topological isomorphism of the space $L_2^a(\mathbb{R}^n)$ onto $A(\Omega_a, L_2(\mathbf{W}))$.

We remind the reader that the space $L_2^a(\mathbb{R}^n)$ consists of functions whose L_2 norms on any shifted compact $K+\gamma$ (where $\gamma \in \Gamma$) decays exponentially with respect to $|\gamma|$, up to $e^{-a|\gamma|}$ (see the precise Definition 3.4.3). The space $A(\Omega_a, L_2(\mathbf{W}))$ consists of $L_2(\mathbf{W})$-valued functions on $\mathbb{T}^*$ that extend analytically to the tubular neighborhood Ω_a of "half-width" a of this torus (see (3.3.4)). Thus, as for the standard Fourier transform, the Floquet one trades analyticity for exponential decay.

The moral is: *If one wants to make the Wannier function* (7.2.1) *decay exponentially, one needs to make sure that the original family of Bloch functions $\phi(k,x)$ is analytic with respect to k in a tubular complex neighborhood Ω_a of the k-torus. The "wider" the analyticity domain is (i.e., the larger a is), the faster is the decay. If analyticity extends to the whole space $(\mathbb{C}\setminus\{0\})^n$, the decay is super-exponential, that is, faster than $e^{-a|\gamma|}$ for* ***any*** *$a > 0$.*

Smoothness conditions that are weaker than analyticity produce slower decaying Wannier functions. We collect all these statements in the following result:

Theorem 7.2.3

(1) *If the Bloch function $\phi(k,x)$ is analytic with respect to the multiplier $z = e^{ik}$ in a (tubular) complex neighborhood Ω_a of the torus $\mathbb{T}^*$, then the corresponding Wannier functions decay exponentially. The "wider" the analyticity domain is, the faster is the decay.*

[5] The classical Euclidean Fourier transform, as we know does exactly the same: trading smoothness for decay.

(2) *C^∞ dependence of $\phi(k.\cdot)$ on k leads to super-polynomial decay, that is, faster than any power of $|k|$.*
(3) *Finite smoothness leads to algebraic decay.*
(4) *The Bloch function belonging to L_2 translates to L_2. That is, the resulting Wannier function $w(x)$ belongs to $L_2(\mathbb{R}^n)$ and*

$$\int_{\mathbb{R}^n} |w(x)|^2 dx = \int_{\mathcal{B}} \|\phi(k,\cdot)\|^2_{L_2(\mathbf{W})} d'k = \int_{\mathbb{T}^*} \|\phi(z,\cdot)\|^2_{L_2(\mathbf{W})} d'z. \quad (7.2.3)$$

(5) *If the Wannier function $w(x)$ belongs to $l_1(L_2(W))$, that is,*

$$\sum_{\gamma\in\Gamma} \|w\|_{L_2(W+\gamma)} < \infty, \quad (7.2.4)$$

this implies continuity of the corresponding Bloch one $\phi(k,\cdot)$ with respect to k.

Proof We have already justified the first statement. The subsequent three statements come from Theorem 3.4.14, while the last is a simple exercise on Fourier series (absolute summability of a Fourier series implies continuity of its sum). □

7.3 Shifts of Wannier Functions

It will be useful to have a Wannier function $w(x)$ whose lattice shifts $w(x-\gamma)$ are mutually orthogonal (even better, ortho-normal), since these shifts are used as parts of a Wannier function basis. To achieve this, one can use the following (immediate to check) formula:

$$w(x+\gamma) = \int_{\mathcal{B}} e^{ik\cdot\gamma} \phi(k,x) d'k \text{ for } \gamma \in \Gamma. \quad (7.3.1)$$

Thus, the γ-shifts of w correspond to the multiplications of the Bloch–Floquet function ϕ by the characters $e^{ik\cdot\gamma}$.

What does it take for the family of shifts of a Wannier function to be orthonormal? The norms of the shifted functions are the same, so the only issue is to make them orthogonal. Here simple answers are provided by the following corollary of Theorems 3.4.1 and 3.4.21:

Lemma 7.3.1

(1) *The shifted Wannier functions $w_\gamma(x) := w_(x-\gamma)$ all have the same L_2 norm.*
(2) *They are mutually orthogonal for $\gamma \in \Gamma$ if and only if the $L_2(\mathbf{W})$-norm of $\phi(k,x)$ in (7.2.1) does not depend on k.*

Proof The first statement, as we have already mentioned, is obvious.

Consider the two shifted version of w: w_{γ_1} and w_{γ_2}, where $\gamma_1 \neq \gamma_2 \in \Gamma$. Then, using (7.3.1), we get

$$w_{\gamma_1} \cdot w_{\gamma_2} = \int_{\mathcal{B}} e^{ik(\gamma_1-\gamma_2)} \|\phi(k)\|^2_{L_2(W)}\, d'k. \tag{7.3.2}$$

The right hand side is recognizable as the $(\gamma_1 - \gamma_2)$th Fourier coefficient of the function $\|\phi(k)\|^2_{L_2(W)}$ on the torus. Thus, all Fourier coefficients corresponding to $\gamma \neq 0$ must vanish, which is equivalent to the function being constant. □

Remark 7.3.2 If the condition of k-independence of the norm in (2) of the Lemma is not satisfied, then the shifts of the corresponding Wannier function are not mutually orthogonal. However, if the norm $\|\phi(k,\cdot)\|_{L_2(\mathbf{W})}$ never vanishes, one can normalize it to

$$\psi(k,\cdot) := \frac{\phi(k,\cdot)}{\|\phi(k,\cdot)\|_{L_2(\mathbf{W})}}$$

to get a new Wannier function

$$w'(x) = \int_{\mathcal{B}} \psi(k,x)d'k,$$

which already has orthonormal shifts, due to Lemma 7.3.1.

Thus, we get the following result:

Theorem 7.3.3 *The set of Wannier functions w_γ is orthonormal if and only if the corresponding Bloch function satisfies the condition*

$$\|\phi(k,\cdot)\|_{L_2(W)} = 1 \textit{ for all } k \in \mathcal{B}. \tag{7.3.3}$$

7.4 Spectral Bands and Composite Wannier Functions: Gapped Case

Let us now think of how one could get appropriate Bloch functions for producing fast decaying Wannier ones. Hold on, we discussed Bloch eigenfunctions as functions of the quasi-momentum in Section 5.5! Let us recall what the outcome was.

Let us pick a band function $\lambda_j(k)$. Theorem 5.5.1 guarantees the possibility of choosing the eigenfunction $\phi_j(k,\cdot)$ as a piecewise analytic L_2-function on $\mathcal{B}$ with values in $L_2(\mathbf{W})$. Alas, as our earlier discussion shows, this is insufficient for the exponential (or even algebraic) decay of the corresponding Wannier function, with issues arising when the bands cross. Indeed, then smooth continuation of $\lambda_j(k)$ and $\phi_j(k,\cdot)$ through the crossing might not be (and usually is not) possible.

Well, what can one do? One expects the situation to be better when the chosen **band is isolated**, that is, does not intersect other bands. In this case, the band function $\lambda_j(k)$ is periodic and analytic with respect to k in a neighborhood of the real space. However, choosing an analytic **and nonvanishing** branch of the eigenfunction might still be impossible. In fact, even continuity of this function with respect to the quasi-momentum might not be achievable. Indeed, finding an analytic and periodic $\phi_j(k,\cdot)$ such that the norm $\|\phi_j(k,\cdot)\|_{L_2(\mathbf{W})}$ never vanishes means finding a non-vanishing analytic section of the 1D[6] holomorphic complex vector bundle

$$\Lambda_j := \bigcup_{\mathcal{D}_a} \mathrm{Ker}(H(k) - \lambda_j(k)I) \tag{7.4.1}$$

over a Γ^*-periodic neighborhood $\mathcal{D}_a$ of the real space.

Remark 7.4.1 Taking into account Γ^*-periodicity and considering Floquet multipliers $z = e^{ik}$ instead of quasi-momenta, one can reinterpret this bundle as one over a neighborhood Ω_a of the unit torus $\mathbb{T}^*$. We adopt this view here, while abusing notations by keeping the same notation for the bundle.

Due to one dimensionality of the bundle, existence of a merely continuous nonvanishing section of Λ_j is equivalent to its triviality. The bundle, however, might be nontrivial, which will be a topological obstruction to existence of not only analytic, but even continuous nonvanishing section (Thouless, 1984). Thus, in particular, due to the last statement of Theorem 7.2.3, would prohibit even slow L_1 decay of the Wannier function in question.

If even a single separate spectral band might be problematic, what can one say when the spectral bands intersect? In general, not much,[7] if for instance there are no spectral gaps (besides the semi-infinite one, omnipresent for semi-bounded operators).

This leads us in Section 7.5 into another rabbit hole.

7.5 Composite Bands and Bloch Bundles: Gapped Case

If at least one finite gap exists in the spectrum, one finds a finite closed part of the spectrum disjoint with the rest of it.

Namely, let $S \subset \sigma(H)$ be a subset consisting of m (intersecting or not) spectral bands and surrounded by spectral gaps (see Fig. 7.1).

Definition 7.5.1 We will call such a subset a **composite band**.

Let us surround S with a contour $\Gamma \subset \mathbb{C}$ not intersecting $\sigma(H)$ (as in Fig. 7.1). The corresponding spectral projector,

$$P_S := \frac{1}{2\pi i} \oint_{\Gamma} (\zeta I - H)^{-1} \, d\zeta \tag{7.5.1}$$

[6] Assuming simplicity of the eigenvalue.

[7] See, however, information of the **metallic Wannier function** in the comments section to this chapter.

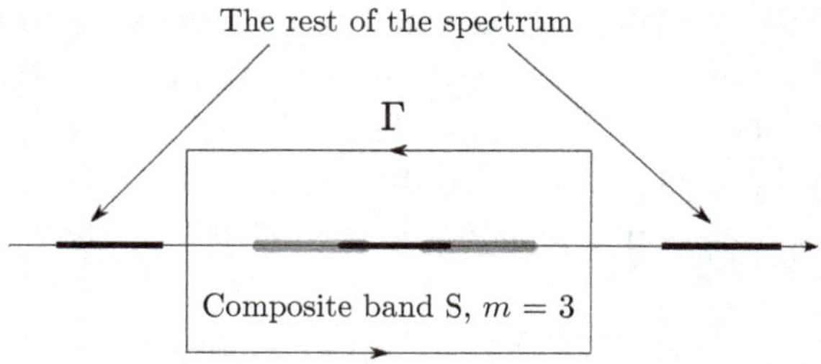

Figure 7.1 A composite band consisting of three spectral bands.

projects orthogonally onto the spectral subspace $\mathcal{E}_S$ of operator H, corresponding to S.

Since the spectra of $H(k)$ are subsets of the whole spectrum $\sigma(H)$, contour Γ also surrounds the corresponding parts of the spectra of all operators $H(k)$. We thus can introduce the corresponding m-dimensional spectral projector for $H(k)$:

$$P(k) := \frac{1}{2\pi i} \oint_{\Gamma} (\zeta I - H(k))^{-1} \, d\zeta. \tag{7.5.2}$$

Projector $P(k)$ depends analytically on k in a complex neighborhood of $\mathbb{T}^*$ in $\mathbb{C}^n$. Thus, its range forms over this neighborhood an analytic vector bundle Λ_S.

Definition 7.5.2 The vector bundle Λ_S is called the **Bloch bundle that corresponds to the composite band** S.

The next auxiliary result follows easily from the direct integral decomposition,

$$H = \int_{\mathbb{T}^*}^{\oplus} H(k). \tag{7.5.3}$$

Lemma 7.5.3 *L_2-sections of the bundle Λ_S form the invariant subspace $\mathcal{E}_S$ for H that corresponds to the separate part $S \subset \sigma(H)$ of the whole spectrum.*

We will now look for a finite set of "nice" sections $\phi_j(k)$, $j = 1, \ldots, l$ of Λ_S that generate the whole space of sections. We then expect that the corresponding Wannier functions and their lattice shifts could provide a basis of $\mathcal{E}_S$ consisting of sufficiently well decaying functions.

Remark 7.5.4 Notice that we are NOT insisting on these sections to correspond to eigenfunctions of the operator. Thus, we will be dealing with **generalized**, or **composite** Wannier functions. As promised, we will drop the word "generalized" in the future discussions.

The natural guess would be to take m sections that form a basis in each fiber.

Well, this requires (and is in fact the definition of the) triviality of the bundle. As in the case when $m = 1$, to make the corresponding Wannier functions decay exponentially, we need these sections to be analytic in a neighborhood of the torus.

Seemingly a simpler question is about the bundle's **topological** triviality, that is, existence of m **continuous basis sections**. Indeed, topological non-triviality would definitely prevent analytic triviality. The question arises whether there are any additional **analytic** obstructions to analytic triviality, besides the topological ones. The following result is a simple incarnation of the (highly nontrivial) so-called **Oka's principle** (see, e.g., Grauert and Remmert, 2004; Gunning and Rossi, 2009 and Appendix A).

Theorem 7.5.5 (e.g., Kuchment, 2009) *Topological triviality of the Bloch bundle Λ_S over $\mathbb{T}^*$ is equivalent to its analytic triviality in a complex neighborhood of this torus.*

Proof Indeed, one can pick a sufficiently narrow neighborhood Ω_a of the torus, which is the product of $1D$ complex annular domains. Then, as any product of $1D$ complex domains, it is a **Stein manifold** (Grauert and Remmert, 2004), and the earlier result is a part of the famous theorem of H. Grauert (1958) for such manifolds. Now, noticing that $\mathbb{T}^*$ is a retract of Ω_a finishes the proof. □

Thus, the question reduces only to the topological triviality of the Bloch bundle over the torus $\mathbb{T}^*$.

Theorem 7.5.6 *Let the Bloch bundle be m-dimensional. Then the following statements are equivalent:*

(1) *The bundle is trivial.*
(2) *There exists a set of m exponentially decaying Wannier functions w_j, $j = 1, \dots m$ such that their shifts $w_{j,\gamma}$, $j = 1, \dots, m$, $\gamma \in \Gamma$ forms an orthonormal basis in the spectral subspace $\mathcal{E}_S$.*
(3) *There exists a set of m Wannier functions w_j, $j = 1, \dots m$ such that $w_j \in l_1(L_2(W))$ and their shifts $w_{j,\gamma}$ for $\gamma \in \Gamma$ forms an orthonormal basis in the spectral subspace $\mathcal{E}_S$.*

Proof (1) $\Rightarrow$ (2) According to Theorem 7.5.5, topological triviality of the bundle implies its analytic triviality. Thus, Λ_S admits m sections $\phi_j(k)$ analytic in a neighborhood of the torus and forming a basis in each fiber. Now, using the "analytic" Gram–Schmidt procedure as described before Theorem 5.6.1, we obtain a basis analytic in a neighborhood of the torus and orthonormal on the torus. The corresponding Wannier functions and their lattice shifts decay exponentially and, according to Lemma 7.3.1, form an orthonormal system.

It remains to prove that this system is complete in the spectral subspace $\mathcal{E}_S$, to which they belong. Indeed, after Floquet transform, the subspace $\mathcal{E}_S$ becomes the space of all vector functions of the form $\sum_j f_j(k)\phi_j(k,\cdot)$, where each $f_j(k)$ is any Γ^*-periodic scalar function such that $\int_{\mathcal{B}} \|f(k)\|^2 d'k < \infty$. On the other hand, finite linear combinations of Wannier functions $\sum\limits_{j,\Gamma} \alpha_{j,\gamma} w_{j,\gamma}$ transform into functions $\left(\sum\limits_{j,\Gamma} \alpha_{j,\gamma} e^{ik\cdot\gamma}\right)\phi_j(k,\cdot)$. Since, according to the standard Fourier series theory (Stein and Weiss, 1971), trigonometric polynomials $\left(\sum\limits_{\Gamma} \alpha_{j,\gamma} e^{ik\cdot\gamma}\right)$ are Γ^*-periodic and dense in $L_2(\mathcal{B})$, this proves completeness of the set of Wannier functions $w_{j,\gamma}$ in $\mathcal{E}_S$.

(2) $\Rightarrow$ (3) This implication is obvious, since exponential decay implies L_1 property.

(3) $\Rightarrow$ (1) An inverse Floquet transform applied to the Wannier functions w_j provides, according to (5) in Theorem 7.2.3, a set of continuous basis sections ϕ_j, and thus the bundle is (topologically, and thus analytically) trivial. □

Remark 7.5.7 Alas, the Bloch bundle Λ_S over the torus $\mathbb{T}^*$ can be indeed topologically nontrivial in physical situations, for example, in the presence of magnetic potential (Thouless, 1984) or in the case of topological insulators (Bernevig, 2013).

So, we have two major questions to answer:

- What are the conditions sufficient for the triviality of the Bloch bundle?
- What can be done when the bundle is nontrivial?

We will address them in the following sections.

7.6 Conditions of Triviality

So, as we have concluded, the existence of a basis (which can be then made orthonormal) of exponentially decaying Wannier functions corresponding to the composite band S disjoint from the rest of the spectrum is equivalent to the triviality of the corresponding Bloch bundle Λ_S. Thus, much effort has been concentrated on establishing sufficient conditions of the triviality, as well as on efficient finding the Wannier bases; see, for example, Auckly and Kuchment (2018); Cornean et al. (2016); Kuchment (2009); Panati (2007); Panati and Pisante (2013): see also Zaidenberg et al. (1975) and references therein, and Appendix D for surveys of techniques of Banach bundles.

There are two important parameters involved: the spatial dimension n, with the most interesting "physical," when $n \leq 3$, and the number m of spectral bands constituting the composite band S. **The 1D case, that is, $n = 1$, presents no issues,** since

Lemma 7.6.1 *There are no nontrivial complex vector bundles over the circle ("1D torus").*

Proof The simple proof of this statement can be recommended as an exercise for an interested reader. It is based upon connectivity of $\mathbb{C} \setminus \{0\}$. □

The things are not that simple in the case when $n > 1$, since then nontrivial bundles over $\mathbb{T}^n$ do exist. There is an important case when these obstructions do not materialize. This is when the operator H commutes with complex conjugation of the functions, that is, has real coefficients.[8] In particular, if $u(x) = e^{ik\cdot x}v(x)$ is a Floquet solution with the quasi-momentum k, then $\overline{u(x)} = e^{-ik\cdot x}\overline{v(x)}$ is a Floquet solution with the quasi-momentum $-k$. Thus, in the Floquet structure of H the symmetry $k \mapsto -k$ is present. In physics, this corresponds to the **time reversal symmetry** and holds for Schrödinger operators with real electric potentials, but breaks down in the presence of a magnetic field.

The following result was proven in Nenciu (1983).

Theorem 7.6.2 *Let the coefficients of the self-adjoint elliptic Γ-periodic operator H be real,[9] and I_j be an isolated band of the spectrum corresponding the band function $\lambda_j(k)$, Then the Bloch bundle Λ_{I_j} is analytically trivial, and thus there exists a corresponding exponentially decaying Wannier function $w_j(x)$ whose Γ-shifts $w_{j,\gamma}(x) = w_j(x-\gamma)$ form an orthonormal basis in the corresponding spectral subspace $\mathcal{E}_{I_j}$.*

Remark 7.6.3 This statement allows arbitrary dimension n, but requires that $m = 1$.

The next result of Panati (2007), under the same symmetry conditions allows any "physical" dimension $n \leq 3$ and arbitrary m (i.e., composite bands are allowed).

Theorem 7.6.4 *Let the coefficients of the self-adjoint elliptic Γ-periodic operator H in $\mathbb{R}^n$, $n \leq 3$ be real and composite band $S \subset \mathbb{R}$ be the union of m spectral bands of H. We again assume that S is separated by gaps from the rest of the spectrum. Then the bundle Λ_S is analytically trivial. Hence, there exist m exponentially decaying generalized Wannier functions $w_j(x)$ whose Γ-shifts $w_{j,\gamma}(x) = w_j(x-\gamma)$ form an orthonormal basis in $\mathcal{E}_S$.*

If a bundle of rank m is trivial, this means that there is a basis of m nonzero continuous sections. Let now it be nontrivial. What about existence then of at least one such (e.g., nonvanishing) section? If it exists, then the bundle has a trivial 1D sub-bundle, and thus splits into the direct sum of this sub-bundle and a bundle of

[8] In fact, some other symmetries suffice, see, for example, Panati (2007).

[9] That is, the time reversal symmetry holds.

rank $m - 1$. Thus, the question reduces to the bundle of smaller dimension. For how long can one can be chipping off the dimension this way? This leads to the following result:

Lemma 7.6.5 *Any complex vector bundle of rank larger than* 1 *over a CW-complex* X *of dimension not exceeding* 3 *admits a nonvanishing section.*

Proof We will construct the section by starting with the cells of dimension zero in the CW-complex and then proceed by induction with respect to the dimension of the skeleton. Clearly one can find a nonzero (in fact unit length, if we introduce a fiber metric on the bundle) section over the (discrete) zero-skeleton of X. Let us now proceed by induction. If there is a section over the k-skeleton, we need to try to extend it to the $(k+1)$-skeleton. By induction hypothesis, we already have a section over the boundary of each $(k+1)$-cell, which is a map from the boundary of the cell to the unit sphere in the fiber: $\mathcal{S}^k \mapsto \mathcal{S}^{2m-1}$. Now, for $m > 1$, the fundamental group and the second homotopy group of S^{2m-1} are trivial, so, due to our restriction on the dimension of X, the section can be extended. □

This leads us to the following statement:

Theorem 7.6.6 *Any complex vector bundle E of rank $m > 1$ over a three-complex splits into the direct sum of a trivial $(m-1)$-dimensional bundle F and a line bundle L:*

$$E = F \oplus L. \tag{7.6.1}$$

In particular,

(1) *$c_1(E) = c_1(L)$, where c_1 is the first Chern class;*
(2) *E is trivial if and only if L is trivial;*
(3) *E is trivial if and only if $c_1(E) = 0$.*

Corollary 7.6.7 *In dimensions $d \leq 3$, if the rank of the Bloch bundle is m, then there exists a set of $m - 1$ orthonormal Wannier functions in the corresponding spectral subspace.*

Thus, only one more function is missing (and might not exist) for a basis.

7.7 Tight (Parseval) Frames of Exponentially Decaying Wannier Functions

Suppose again that S is a multiple band S consisting of m bands of the spectrum, separated from the rest of the spectrum by gaps. Consider the corresponding m-dimensional analytic Bloch bundle Λ_S over $\mathbb{T}^*$ (see Definition 7.5.2). If it is topologically (and thus, by Oka's principle also analytically, see Theorem 7.5.5) trivial, it has an analytic basis of Bloch functions $\phi_1(k,\cdot),\dots,\phi_m(k,\cdot)$, which can be assumed ortho-normal over the torus. According to Theorem 7.5.6, the corresponding exponentially decaying Wannier functions

$$w_j = \int_{\mathbb{T}^*} \phi_j(k,\cdot)dk \tag{7.7.1}$$

and their Γ-shifts form an ortho-normal basis in the spectral subspace $\mathcal{E}_S$ of the operator H that corresponds to the isolated part of the spectrum S.

This construction does not work, if the Bloch bundle is nontrivial. For a while, presence of this obstacle (bundle non-triviality) had not been realized, and thus there might had been a belief that nice Wannier bases always exist. Alas, Thouless (1984) showed that in the presence of magnetic terms in a periodic Schrödinger operator, the Bloch bundle can be nontrivial (non-triviality of the Bloch bundle also arises in topological insulators theory [Bernevig, 2013], where it is the main desirable factor).

It seems that nothing can be done. Indeed, according to Theorem 7.5.6, one should give up the hope of finding bases of even very moderately decaying Wannier functions. So, what can be done? One of suggestions is to try to find more than m (and thus linearly dependent) fast decaying functions such that they and their Γ-shifts "nicely" generate the spectral subspace (although being linearly dependent). What could be a "nice" overdetermined analog of an orthonormal basis? An option, popular in the recent decades, is given by the so-called **frames**, especially the so called **tight (or Parseval) frames** (see, e.g., Han et al., 2007 and references therein).

7.7.1 Frames

An overdetermined system of vectors cannot form a basis or be orthogonal. However, there are the following notions (e.g., Han et al., 2007) that replace the basis property and orthonormality in the overdetermined case.

Definition 7.7.1

- A set of vectors v_j, $j = 1,2,\dots$ in a Hilbert space H with the inner product $(\cdot,\cdot)$ is said to be a **frame**, if for some constants $A, B > 0$ and any vector $u \in H$, the following inequality holds:

$$A\sum_j |(u,v_j)|^2 \le \|u\|^2 \le B\sum_j |(u,v_j)|^2.$$

- This set is said to be a **tight (or Parseval) frame** if $A, B = 1$, that is, for any vector $u \in H$ the equality holds

$$\sum_j |(u,v_j)|^2 = \|u\|^2.$$

Remark 7.7.2

- It is known (Han et al., 2007) that tight frames are precisely the orthogonal projections of orthonormal bases from larger Hilbert spaces.
- A wonderful property of any tight frame $\{w_j\}$ is that the formulas for expanding an arbitrary vector into it (Fourier analysis) and reassembling the vector from the resulting coefficients (Fourier synthesis) are identical to those that one has for expanding into orthonormal bases. This explains the name Parseval frame.

7.7.2 Wannier Tight Frames

The next simple result shows that one can keep exponential decay, increasing the number of Wannier functions and sacrificing orthogonality, while still having a complete system of functions.

Theorem 7.7.3 *Let H be a self-adjoint elliptic Γ-periodic operator in $\mathbb{R}^n$, $n = 1, 2, \ldots$ and $S \subset \mathbb{R}$ be the union of m spectral bands of H separated from the rest of the spectrum. Then there exist a (finite) number $l \geq m$ of exponentially decaying generalized Wannier functions w_j, $j = 1, \ldots, l$ such that their shifts $w_{j,\gamma}$ for $\gamma \in \Gamma$ form a complete system in the spectral subspace $\mathcal{E}_S$ (i.e., the set of all their finite linear combinations is dense in the space).*

Proof We will prove this statement in an equivalent formulation: there exists a finite number of analytic sections $\phi_j(k, \cdot)$, $j = 1, \ldots, l$ of the bundle Λ_S over Ω_a for some $a > 0$ such that the sections $\phi_j(k, \cdot) := e^{ik\cdot\gamma}\phi_j(k, \cdot)$ for $j = 1, \ldots, l, \gamma \in \Gamma$ form a complete set of functions in the space of L_2-sections of the bundle Λ_S.

It is a simple exercise to show that it is sufficient for that to have a finite family of analytic sections ϕ_j such that they span the whole fiber of the bundle at each point $z \in \mathbb{T}^*$. Recall now that the bundle Λ_S is trivial locally, that is, in a neighborhood of each point. Let us fix two finite trivializing covers U_j and $W_j \Subset U_j$, $j = 1, \ldots, l$ of $\mathbb{T}^*$ by complex domains such that their union is a Stein domain (e.g., the union can be made equal to Ω_a for some $a > 0$). Then on each U_j, by assumption, there is an analytic basis $\psi_{j,t}(k)$, $t = 1, \ldots, m$ in the restriction of the bundle Λ_S onto U_j. Due to the Stein property, these holomorphic sections can be approximated with any precision uniformly on $\overline{W_j}$ by global (i.e., defined on the whole Ω_a) analytic sections $\phi_{j,t}(k)$ (see, e.g., Ch. VII of Hörmander, 1990 or Theorem 1.5.9 of Kuchment, 1993). Doing this for each j, with a sufficiently close approximation, one achieves the desired property for the functions $\phi_{j,t}$. □

In this result, we seemingly have sacrificed too much. We not only have an excessive (in comparison with the number of spectral bands involved) number of Wannier functions, we lost the orthonormal basis property, which was replaced by a much weaker (and much less useful) completeness. Well, one cannot help it having more than m and non-orthogonal Wannier functions, as the following statement holds:

Proposition 7.7.4 *If in the previous theorem either the number of Wannier functions is exactly equal to m, or the shifts of the Wannier functions w_j form an orthonormal basis, then the bundle Λ_S is trivial.*

Proof Indeed, if $l = m$, one notices that in this case the Floquet transforms ϕ_j of the Wannier functions must be analytic and at each point span the whole fiber of the bundle. Since the fibers have dimension m, we conclude that ϕ_j must form an analytic basis of the bundle, and thus the bundle is trivial.

Suppose now that the functions $w_{j,\gamma}$ from the preceding theorem are orthonormal. Normalization property, as we already know, means that the functions $\phi_j(k,\cdot)$ have a constant norm in $L_2(\mathbf{W})$. Checking orthogonality of $w_{j_1,0}$ to $w_{j_2,\gamma}$ for $j_1 \neq j_2$ and arbitrary $\gamma \in \Gamma$, one obtains that

$$\int_{\mathbb{T}^*} e^{ik\cdot\gamma}(\phi_{j_1}(k,\cdot),\phi_{j_2}(k,\cdot))_{L_2(\mathbf{W})}\,dk = 0,$$

for all $\gamma \in \Gamma$. This means that the whole Fourier series with respect to k of the function $(\phi_{j_1}(k,\cdot),\phi_{j_2}(k,\cdot))_{L_2(\mathbf{W})}$ on $\mathbb{T}^*$ vanishes. Thus, $(\phi_{j_1}(k,\cdot),\phi_{j_2}(k,\cdot))_{L_2(\mathbf{W})}$ is equal to zero almost everywhere, and hence the functions ϕ_{j_1}, ϕ_{j_2} are orthogonal for each k. Since the dimension of the fiber is m, we conclude that the number of the Wannier functions does not exceed m, and thus, according to the previous case, the bundle is trivial. □

So, if one cannot help with excessive number of Wannier functions, can one improve on the completeness statement? The answer is a "yes," as the next result shows.

Theorem 7.7.5 *Let H be a self-adjoint elliptic Γ-periodic operator in $\mathbb{R}^n$, $n = 1, 2, \ldots$ and $S \subset \mathbb{R}$ be the union of m spectral bands of H. Suppose that S is separated from the rest of the spectrum. Then there exists a finite number $l \geq m$ of exponentially decaying composite Wannier functions $w_j(x)$ such that their shifts $w_{j,\gamma} := w_j(x-\gamma), \gamma \in \Gamma$ form a tight frame in the spectral subspace $\mathcal{E}_S$ of operator H. This, in particular, means that for any $f(x) \in \mathcal{E}_S$, the equality holds*

$$\int_{\mathbb{R}^n} |f(x)|^2 dx = \sum_{j,\gamma} \left| \int_{\mathbb{R}^n} f(x)\overline{w_{j,\gamma}(x)}\,dx \right|^2. \tag{7.7.2}$$

Here the number l is the smallest dimension of a trivial bundle containing an equivalent copy of Λ_S. In particular, $l = m$ if and only if the bundle Λ_S is trivial, in which case an orthonormal basis of exponentially decaying composite Wannier functions exists, as we know already.

Proof Let us start with establishing first the following auxiliary statement:

Lemma 7.7.6 *There exists a finite-dimensional analytic sub-bundle Λ' of the trivial bundle $\Omega_a \times L_2(\mathbf{W})$, such that*

(1) *The fibers at the same point k of Λ_S and Λ' intersect only at zero.*
(2) *The analytic sub-bundle $\Phi := \Lambda_S \oplus \Lambda'$ in $\Omega_a \times L_2(\mathbf{W})$ is trivial.*

There are several ways to show that this lemma holds. For instance, let us choose an abstract finite-dimensional analytic bundle Λ' such that added to Λ_S restricted to some $\overline{\Omega_b}$ (where $\overline{\Omega}$ is the closure of Ω) it leads to a trivial bundle of dimension l. This is possible in the topological category (Atiyah, 1989; Husemoller, 1975) and thus, according to Grauert's theorem (Grauert, 1958) and Stein property of Ω_a, also in analytic category.

Remark 7.7.7 Using Gaussian maps, one can make sure (e.g., Proposition 3.5.8 in Husemoller, 1975) that the dimension l of the direct sum $\Lambda_S \oplus \Lambda'$ does not exceed km, where m is the dimension of the fibers of the bundle Λ_S and k is its type, that is, the number of open sets covering its base, over which the bundle is trivial. In the case of an n-dimensional torus as the base, it is easy to see that $k \leq 2^n$. Thus, one can always achieve $l \leq 2^n m$. As we will see later, this upper estimate from Kuchment (2009) is an **awful overshot**.

Now let us consider a complementary to Λ_S infinite-dimensional sub-bundle $\mathcal{F}$ in $\Omega_a \times L_2(\mathbf{W})$, which always exists due to Theorem 3.11 in Zaidenberg et al. (1975), and in our case can be constructed using the spectral projectors. The claim is now that there is an analytic sub-bundle in $\mathcal{F}$ that is isomorphic to Λ'. Indeed, taking the direct sum of Λ' and of the trivial bundle with an infinite-dimensional separable Hilbert space G as a fiber, one gets a Hilbert analytic fiber-bundle over a Stein manifold. According to Kuiper's theorem (Atiyah, 1989; Kuiper, 1965), this bundle is topologically trivial. Now, Bungart's theorem (Bungart, 1967) (an infinite-dimensional version of the theorem of Grauert about fiber bundles over Stein spaces) implies that this sum is also analytically trivial. Since, again by the Kuiper and Bungart theorems, all analytic infinite-dimensional Hilbert bundles with separable fibers over Stein bases are equivalent, we conclude that the direct sum is equivalent to $\mathcal{F}$, which embeds a copy of Λ' into $\mathcal{F}$. This proves the lemma.

Let us now prove the statement of the theorem, using the sub-bundle Λ' as a tool. Consider the trivial bundle $\Phi := \Lambda_S \oplus \Lambda' \subset \Omega_a \times L_2(\mathbf{W})$. Due to its triviality, one can find a basis of its analytic sections $\psi_j(k)$, $j = 1, \ldots, l$. Using the analytic version of the Gram–Schmidt orthogonalization method as described before Theorem 5.6.1, we can produce another basis family, which is orthonormal for real values of k. We can thus assume that $\psi_j(k)$ is already like that. Applying now the spectral projector $P_S(k)$ to this family of sections, we obtain an analytic family of sections $\phi_j(k)$ of the bundle Λ_S, which in each fiber form a tight frame (as the orthogonal projection of an orthonormal basis in a larger space). Consider the two families of generalized Wannier functions: v_j that correspond to ψ_j and w_j that correspond to ϕ_j. The functions $v_{j,\gamma}$ do not necessarily belong to the spectral subspace $\mathcal{E}_S$, but according to Theorem 3.4.1 and Corollary 7.3.1, they form an orthonormal basis in a Hilbert subspace H' that is larger than $\mathcal{E}_S$. (This is the subspace that, after the Bloch–Floquet transform, produces the space of all L_2-sections of the bundle Φ over the torus.) Thus, the family $w_{j,\gamma}$, as the orthogonal projection of $v_{j,\gamma}$ from H' onto $\mathcal{E}_S$, form a tight frame in $\mathcal{E}_S$. □

7.7.3 How Many Additional Wannier Functions Are Needed to Form a Frame?

We discuss here the question of how redundant the Wannier frame needs to be? In other words, how much larger the number l in the previous results must be than the rank (dimension of the fiber) m? The upper bound in Kuchment (2009) was a huge overkill, not helpful in practice. In fact, in small (physical) dimensions the answer is much better.

Theorem 7.7.8 *If $n \leq 3$, a set of $l = m + 1$ of Wannier functions and their Γ-shifts suffice.*

Proof Indeed, let us use Theorem 7.6.6 to expand the bundle Λ_S as follows:

$$\Lambda_S = F \oplus L, \tag{7.7.3}$$

where the bundle F is trivial, and L is a line- (i.e., 1D) bundle. Let L^* be the **dual bundle** to L. Then the bundle $L^* \oplus L$, and thus also $\Lambda_S \oplus L^*$ are trivial. □

Remark 7.7.9 (Auckly and Kuchment, 2018) When $n = 4$, $m + 2$ Wannier functions might be needed.

A better estimate, which also holds in all dimensions and the arbitrary base has been obtained recently in Ballas et al. (2023, Theorem 4.2):

Theorem 7.7.10 *Let $E \mapsto M$ be a rank m complex vector bundle over a n-dimensional smooth (real) manifold. Suppose that $E \mapsto M$ is equipped with a Hermitian structure. Then there exists a Parseval frame for $E \mapsto M$ consisting of $(\lfloor n/2 \rfloor + m)$ sections.*

One notices that this covers the previously mentioned estimates for 2D and 3D tori.

7.8 Not Gapped (Metallic, Intertwined) Case

Up to this point, we have assumed that the bands of interest were surrounded by gaps. The bottom gap is usually not an issue, if the operator is bounded from below, but the upper ones not necessarily exist, at least in the location of interest (for instance, in the metallic case at the Fermi level). In other words, the nth band overlaps with the next one, and more importantly the nth band function $\lambda_n(k)$ might not be smooth, due to intersection with the next one λ_{n+1}. Then the techniques of this chapter are not applicable.[10]

One might expect that some success might still be expected, although one does not hope for equivalence of exponential and faster decays anymore, due to the absence of Oka/Bungart theorems. However, one still can hope to get slower (algebraically or even super-algebraically) decaying Wannier bases/frames. And indeed, this turns out to be possible, if one involves the next, $(n+1)$th band (see, e.g., Cornean et al., 2019a and references therein).

7.9 Comments and References

- **The existence of Wannier bases:** According to Theorem 7.5.6 (proven in Kuchment, 2009), the existence of Wannier bases implies (and, in fact, is equivalent to) triviality of the corresponding Bloch bundle and thus vanishing of the corresponding topological invariants (Chern numbers). This is what makes the existence of such bases not interesting to topological insulators. However, tight frames exist even in topologically nontrivial situations.

  It is also shown in that theorem that existence of a basis consisting of a rather weakly (essentially L_1) decaying Wannier functions implies existence of an exponentially decaying one.

  Later, similar theorems about both statements started appearing, (some requiring stronger decay than in (3) of Theorem 7.5.6), without acknowledging the results of Kuchment (2009). Well, at least the correctness of these was confirmed again:).
- **Non-triviality:** It is interesting to note that while in this chapter we consider non-triviality as an obstacle for existence of bases of fast decaying Wannier functions, it (non-trivially) happens to be a blessing for existence of the magical topological insulators (see, e.g., Bernevig, 2013; Hasan and Kane, 2010; Kitaev, 2009; Prodan and Schulz-Baldes, 2016).
- **Gaps:** We emphasize again that in the case of absence of a gap exponentially decaying Wannier functions do not exist, but ones of slower decay might. For instance, even at the Dirac points of graphene, one can have success (Monaco and Panati, 2014).

[10] However, in the absence of intersection, that is, existence of non-omnidirectional gap, the techniques do apply.

- **Wannier bases:** The original proofs of results on the existence of Wannier bases or Parseval frames (especially in the composite cases) were not very constructive, which diminished their usefulness. More constructive proofs have also appeared (see, for example, Cornean et al., 2016, 2017).
- **Nonperiodicity:** Wannier bases for gapped **nonperiodic** cases have also being considered, see for example, Marcelli et al. (2023); Moscolari and Panati (2023); Prodan (2015), among others.
- **Zak Transforms:** We considered the situation of rather general periodic coefficient elliptic operator, including the case of **periodic magnetic potentials** (see, e.g., Cornean et al., 2019b). However, if a constant magnetic field is present, instead of the lattice of shifts a larger invariance group arises of the so-called **Zak transforms**. If the magnetic flux through the unit cell of periodicity is rational, this group turns out to be commutative, and most of the theory translates to this case, while irrational fluxes essentially destroy the techniques that we use in this book.
- **Noncommutativity:** Handling noncommutative periodicity situations requires drastically different techniques and produces different results, see for instance (Bourne and Mesland, 2021; Ludewig and Thiang, 2020). We do not want to touch it here even with a ten-foot pole, but plan to address it somewhat (at least in a brief survey) in Kuchment (2027).
- **Apology:** Finally, the current state of research on Wannier bases/frames, even in the commutative situation, is vast and the author makes no claim for the earlier discussion to be anything like a complete survey of its state.

8

Operators on Abelian Coverings of Compact Manifolds

8.1 Introduction

In this tiny chapter, we briefly mention the case of **periodic operators on abelian coverings of compact manifolds**.

As will be indicated later and addressed in more detail in Kuchment (2027), the non-abelian case is completely different, unless there is a commutative subgroup of a finite index.

Consider a (noncompact) connected d-dimensional Riemannian manifold N with free, cocompact, and isometric action of a finitely generated discrete group Γ that has a free abelian subgroup of a finite index. Then the quotient space

$$M := N/\Gamma \tag{8.1.1}$$

is compact and inherits the Riemannian structure. We thus have a covering

$$N \overset{\Gamma}{\mapsto} M, \tag{8.1.2}$$

which is analog of the previously used covering $\mathbb{R}^n \overset{\mathbb{Z}^n}{\mapsto} \mathbb{T}^n$.

Remark 8.1.1 Notice that unlike the flat case considered before, the dimensions d of the manifold and the free rank n of the group do not have to be equal. **In many cases, the rank n plays the leading role.**

Since Γ has a finite index free abelian subgroup, then for all our future purposes, we can switch to a finite sub-cover and assume the group Γ to be free abelian, that is, $\Gamma = \mathbb{Z}^n$. Thus, we will assume that we deal with an abelian covering.

8.2 Floquet Techniques

The Floquet theory techniques for Γ-periodic operators on N described in Chapters 3 and 4 survive completely, with the obvious exception of the explicit formulas for the symbols of differential operators $H(k)$ in Section 3.6. However, the author is not aware of results on irreducibility and algebraicity of Fermi–Bloch varieties in this more general setting.

8.3 Spectral Structure

The results concerning general band-gap structure of the spectra of periodic elliptic operators, absence of the singular continuous spectrum, spectral decomposition, and the Schnol's theorem still hold with the same proofs (see, e.g., Kobayashi et al., 1989; Sunada, 1990, 1992 for the analogies with the "flat" case).

However, the situation with absolute continuity of the spectrum, that is, absence of the pure point spectrum (bounded states) becomes not clear, even for "free" Laplace–Beltrami operators on N.

The wonderful Friedlander Theorem 5.7.12, which asserts the absolute continuity under the condition of evenness of the operator with respect to one of the generators of the group $\Gamma = \mathbb{Z}^n$ still holds. The remarkable fact is that while this result was formally proven in Friedlander (2002) in the flat situation only, the proof does not require any modifications for the more general case considered here.

It is also shown in Sunada (1990) that the absolute continuity holds true if the base M admits a nontrivial $\mathbb{S}^1$-action with a parallel generating vector field.

8.4 Spectral Gaps

The question about existence and number of spectral gaps becomes significantly different, even in the case of the (clearly periodic)"free" Laplace–Beltrami operator L. In the flat case considered in Chapter 5, the free operator (Laplacian) has no spectral gaps, its spectral bands overlap, and for its perturbations it is expected (and proven in many cases) that the BCS conjecture on finiteness of the number of gaps holds. It was also mentioned there that a periodic Laplace–Beltrami operator can have an arbitrarily large number of spectral gaps, while the validity of the BCS conjecture is unclear.

One can find in Hempel and Post (2003); Lledó and Post (2007, 2008); Post (2003) various approaches (including high-contrast ones) to and examples of creating spectral gaps on abelian coverings of compact manifolds. The reason is that periodic Laplace–Beltrami operators are not sufficiently "free": one can create various "bumps" of "traps" in the fundamental domain. To avoid this, one can try to assume that the base M is in some sense "homogeneous," for instance having a constant curvature.[1]

[1] Clearly, the interesting case is when the curvature is negative.

Consider the standard covering $\mathbb{R}^n \mapsto \mathbb{T}\,(=\mathbb{R}^n/\mathbb{Z}^n)$. Its deck group is $\mathbb{Z}^n$, which coincides with $H_1(\mathbb{T},\mathbb{Z})$. It can also be understood as the quotient of the universal cover of the torus by the commutator subgroup of the fundamental group $\pi_1(\mathbb{T})$.

Definition 8.4.1 The **maximal abelian covering** of a compact manifold X is the covering $Y \mapsto X$ with the deck group $H_1(X,\mathbb{Z})$, which can be obtained as the quotient of the universal cover of X by the commutator subgroup of the fundamental group $\pi_1(X)$.

The following monotonicity result holds (see, e.g., Sunada, 1985):

Theorem 8.4.2 *Let $Y \mapsto X$ be a Riemannian covering with an amenable (for instance, abelian) deck group. Then*

$$\sigma(-\Delta_X) \subset \sigma(-\Delta_Y),$$

where Δ_M denotes the Laplace–Beltrami operator on a Riemannian manifold M.

Thus, the spectrum of the Laplacian on the maximal abelian covering is the largest (and thus has the fewest gaps) among all abelian coverings of the same base. The following, still not proven, **no gap conjecture** has been formulated by Sunada (1990, 2008):

Conjecture 8.4.3 *Let $Y \mapsto X$ be the maximal abelian covering of a compact Riemannian manifold X of a constant negative curvature. Then the spectrum of the Laplace–Beltrami operator $-\Delta_Y$ on Y has no gaps.*

Remark 8.4.4 Here the constant curvature condition should preclude the possibility of creating "bumps" and "traps" that could lead to gap opening.

8.5 Comments and References

A much richer and difficult questions arise for coverings with non-abelian deck groups. It is planned to address this in Kuchment (2027). Meanwhile, for more information and further references, the reader can browse the following rather random (and incomplete) collection of articles: Bourne and Mesland (2021); Brooks (1981, 1986); Brüning and Sunada (1992a,b); Gruber (2001, 2003, 2000); Lin and Pinchover (1994); Ludewig and Thiang (2020); Sunada (1992, 1988); Zelditch (1990).

Appendix A
Some Information from Complex Analysis

A.1 Several Complex Variables

In some instances of this book, we needed to look at functions (including operator-valued ones) analytically depending on a multidimensional (or even infinitely dimensional) complex parameter. We thus introduce first a class of complex domains/manifolds/analytic sets that is the easiest to handle and which we encounter in this text. These are the so-called **Stein manifolds**. In particular, the so-called **Oka's principle** holds for these manifolds, which, to put it crudely, says that everything that can be done topologically can be done analytically (e.g., topologically trivial vector bundles are automatically analytically trivial). One can find this and other basic notions and results of the several complex variables theory in many books, for instance in Grauert and Remmert (2004); Grauert (1958); Gunning and Rossi (2009); Hörmander (1990); Malgrange (1984).

Definition A.1 A complex analytic countable at infinity manifold[1] X is called a **Stein manifold** if:

(1) it is holomorphically convex (i.e., the holomorphic convex hull of any compact subset of X is compact),
(2) it is holomorphically separable (i.e., for any two distinct points $x, y \in X$ there is a holomorphic function f on X with the property that $f(x) \neq f(y)$,
(3) for any point $x_0 \in X$ there is a set of holomorphic functions on X forming local coordinates in a neighborhood of x_0.

There are several frequently encountered cases, when X is Stein, for example:

- Any 1D domain $X \subset \mathbb{C}$ is Stein. Thus, when dealing with functions of one variable, one should not worry about satisfying the Stein properties.
- Any direct product of Stein manifolds $U_1 \times U_2 \times \cdots \times U_k$ is Stein.

[1] There are also parallel notions for the more general case of analytic sets.

- Any analytic submanifold of $\mathbb{C}^n$ (or even a finite-dimensional analytic submanifold of an infinite-dimensional complex Banach space) is Stein.

A.2 Analytic Sets

Definition A.2 A subset X of an analytic manifold M is said to be **analytic set**, if it can be locally described as the set of common zeros of several analytic functions. If (locally) one such function suffices, the set is **principal analytic**.

We now provide a lemma that collects several basic facts about analytic sets that we will need later. The reader can find them in many books on several complex variables. In particular, all these statements are proven in Chirka (1989, Sections 2.3, 5.3, 5.4, 5.5, Chapter 1) (see also Hörmander, 1990). We start first with a definition.

Definition A.3 A point of an analytic set $A \subset \mathbb{C}^m$ is said to be **regular**, if in a neighborhood of this point the set A can be represented as an analytic submanifold of $\mathbb{C}^m$. The set of all regular points of A is denoted by $\operatorname{reg} A$.

Lemma A.4 *Let A be an analytic set.*

(a) *The set* $\operatorname{reg} A$ *is dense in A. Its complement in A is closed and nowhere dense in A.*
(b) *The set A can be represented as a (maybe infinite) locally finite union of irreducible subsets*

$$A = \bigcup_i A_i$$

called its ***irreducible components****.*
(c) *Irreducible components are closures of connected components of* $\operatorname{reg} A$. *In particular, set A is irreducible if and only if* $\operatorname{reg} A$ *is connected.*
(d) *Let A be irreducible and A_1 be another analytic set such that $A \cap A_1$ contains a non-empty open portion of A. Then $A \subset A_1$. In particular, if f is an analytic function that vanishes on an open portion of A, then f vanishes on A.*
(e) *Any analytic set A has a stratification $A = \bigcup A_j$ into disjoint complex analytic manifolds (strata) A_j such that the union $\bigcup A_j$ is locally finite, the closure $\overline{A_j}$ of each A_j and its boundary $\overline{A_j} \setminus A_j$ are analytic subsets, and such that if the intersection $A_j \cap \overline{A_k}$ of two different strata is not empty, then $A_j \subset \overline{A_k}$ and* $\dim A_j < \dim A_k$.

We will need the following simple corollary (a uniqueness of analytic continuation statement) of this lemma.

Corollary A.5 *Let A be an irreducible proper analytic subset of $\mathbb{C}^n$ such that the intersection $A \cap \mathbb{R}^n$ contains an open part of a smooth $(n-1)$-dimensional submanifold $M \subset \mathbb{R}^n$. If f is an analytic function in a complex neighborhood of M such that $f = 0$ on M, then $f = 0$ on an open subset of A. In particular, if $\mathcal{B}$ is another analytic subset of $\mathbb{C}^n$ such that $M \subset \mathcal{B}$, then $A \subset \mathcal{B}$.*

Proof Considering stratification $A = \bigcup A_j$ (see the last statement in the preceding lemma), one can conclude that only strata of dimension $(n-1)$ can contain open pieces of M. In fact, if there is an A_j of complex dimension at most $(n-2)$ containing an open piece of M, we get the contradiction as follows. Consider a point of M and the tangent space to A_j at this point. This is a complex linear subspace of complex dimension at most $n-2$, which contains a real subspace (tangent space to M) of real dimension $n-1$. This is obviously impossible. So, now we can assume that instead of A we are dealing with one of its strata of dimension $(n-1)$, that is, we may assume that A is smooth. In appropriate analytic coordinates in a complex neighborhood of a point of M one can represent A locally as a complex hyperplane with the real part M. Then standard uniqueness theorem for analytic functions implies that any function f analytic on A and vanishing on M is identically equal to zero on A. If now $\mathcal{B}$ is an analytic subset of $\mathbb{C}^n$ such that $M \subset \mathcal{B}$, then each of the analytic functions locally defining $\mathcal{B}$ has the properties of the function f given earlier. Hence, we conclude that $\mathcal{B}$ contains an open part of A. Then statement (d) of the lemma implies that $A \subset \mathcal{B}$. This finishes the proof of the corollary. □

A stratification analogous to the one described in the last statement of the lemma also holds for real analytic sets (Narasimhan, 1966).

Appendix B

Some Information from Functional Analysis and Operator Theory

In this chapter, we collect some notions and facts from functional analysis that arise in the main text of the book about some classes of linear operators, including in particular Fredholm operators, their indexes, Shatten–Neumann operator classes, and regularized determinants.

B.1 Some Notations Concerning Linear Operators

Let E and F be Banach spaces. We denote by $L(E, F)$ the space of all bounded linear operators $A : E \mapsto F$, equipped with the operator norm. When $E = F$, we use the notation $L(E) := L(E, E)$. The identity operator in E is denoted as I_E (we also use the notations id or I when the domain of the mapping is clear from the context).

We denote by $\mathrm{GL}(E) \subset L(E)$ the group of invertible bounded linear operators acting in E and by $C(E, F) \subset L(E, F)$ the space of compact linear operators acting from E to F (we use $C(E)$ instead of $C(E, E)$).

For a linear operator A acting from E to F we denote by $\mathrm{Ran}\, A$, $\mathrm{Ker}\, A$, $\mathrm{coKer}\, A$, and $\mathrm{rank}\, A$ correspondingly the range, the kernel, the co-kernel, and the rank of A:

$$\begin{aligned}
&\mathrm{Ran}\, A = \{y \in F \mid y = Ax, \text{ for some } x \in E\} \subset F,\\
&\mathrm{Ker}\, A = \{x \in E \mid Ax = 0\} \subset E,\\
&\mathrm{coKer}\, A = \{f \in F^* \mid f(Ax) = 0 \text{ for all } x \in E\} \subset F^*,\\
&\mathrm{rank}\, A = \dim \mathrm{Ran}\, A.
\end{aligned}$$

B.2 Spectral Theorem and Spectrum Classification

One can find the information below in many books on spectral theory, for instance a good reference for this part is Reed and Simon (1972, vol. 1, Ch. VII and VIII).

There are several possible classifications of the types of spectrum; we review here the classes that are most important to us.

Definition B.1

- A point $\lambda \in \mathbb{C}$ is said to be an **eigenvalue** of A, if the kernel of the operator $A - \lambda I$ is non-trivial. The elements of this kernel are said to be the **eigenfunctions** corresponding to the **eigenvalue** λ.
- The set of all eigenvalues is called the **point spectrum** of the operator A and is denoted $\sigma_p(A)$.
- The set of all isolated eigenvalues of finite multiplicity is the **discrete spectrum** $\sigma_d(A)$.
- The complement to the discrete spectrum in the whole spectrum is called the **essential spectrum** $\sigma_{ess}(A) := \sigma(A) \setminus \sigma_d(A)$.
- If operator $(A - \lambda I)$ is injective, but its range $Ran(A - \lambda I)$ is not dense in H, then λ is in the **residual spectrum** .

If the operator is self-adjoint, there are several simplifications. In particular, the spectrum is real, the residual spectrum is empty and the essential spectrum can be characterized as the set of all complex numbers λ such that $(A - \lambda I)$ is not a Fredholm operator (see later in this Chapter). Moreover, the spectral theory of self-adjoint operators gets more detailed, as described below.

Let A be a (bounded or unbounded) self-adjoint operator in H. Here is one (functional calculus form) version of the spectral theorem for self-adjoint operators.

Theorem B.2 (Theorem VIII.5 of Reed and Simon, 1972) *Let A be a self-adjoint operator in H. There exists a unique map ϕ from the space of bounded Borel functions on $\mathbb{R}$ to the space $L(H)$ with the following properties:*

(1) *ϕ is an algebraic $*$-homomorphism of $*$-algebras.*
(2) *$\|f(A)\| \leq \|f\|_{L^\infty}$, where we denote $f(A) := \phi(f) \in L(H)$.*
(3) *Let $f_n(x)$ be such that*

$$f_n(x) \underset{n\to\infty}{\longrightarrow} x \text{ for all } x$$

and $|f_n(x)| \leq |x|$. Then for any w in the domain $\mathcal{D}(A)$,

$$\lim_{n\to\infty} \phi(f_n)w = Aw.$$

(4) *If f_n are uniformly bounded and $f_n(x) \to f(x)$ for any x, then $f_n(A) := \phi(f_n)$ strongly converges to $f(A) := \phi(f)$.*
Moreover,
(5) *If $Aw = \lambda w$, then $f(A)w = f(\lambda)w$.*
(6) *If $f \geq 0$, then $f(A) \geq 0$.*

Remark B.3 Additionally,

- Let $\lambda \notin \sigma(A)$ and $r_\lambda(z) := (\lambda - z)^{-1}$, then $r_\lambda(A) = (\lambda I - A)^{-1}$ (the **resolvent**).
- If $\operatorname{supp} f \cap \sigma(A) = \emptyset$, then $f(A) = 0$.

Let now $w \in H$. Then one can define a measure μ_w on the real line as follows:

$$\int_{\mathbb{R}} f(z) d\mu_w(z) := (w, f(A)w).$$

The standard decomposition of measures allows one to split this measure on $\mathbb{R}$ into three parts: pure point, absolutely continuous, and singular continuous with respect to the Lebesgue measure. One can define now the following invariant subspaces for the operator A:

$$\begin{aligned} H_{\text{pp}} &:= \{w|\mu_w \text{ is purely point}\}, \\ H_{\text{ac}} &:= \{w|\mu_w \text{ is absolutely continuous w.r.t. Lebesgue measure}\}, \\ H_{\text{sc}} &:= \{w|\mu_w \text{ is singular continuous w.r.t. Lebesgue measure}\}. \end{aligned}$$

We can now introduce the classification of the spectrum of A into **pure point** $\sigma_{\text{pp}}(A)$, **absolutely continuous** $\sigma_{\text{ac}}(A)$, and **singular continuous** $\sigma_{\text{sc}}(A)$ as the spectra of the restrictions of the operator A onto the invariant subspace H_{pp}, H_{ac}, and H_{sc} correspondingly.

The reader is referred to Akhiezer and Glazman (1993); Cycon et al. (1987); Hislop and Sigal (1996); Reed and Simon (1972) concerning deeper details of spectral theory, including **spectral measure** and **density of states**.

B.3 Lifting and Open Mapping Theorems

In Banach spaces that are not isomorphic to Hilbert ones there are closed subspaces without a closed complementary subspace. In other words, one can have a continuous surjective linear mapping $A : E \mapsto F$ between Banach spaces, which does not have a continuous linear right inverse $B : F \mapsto E$. This section provides a tool that sometimes helps to overcome this hurdle.

Let $N \subset E$ a closed linear subspace in a Banach space E and E/N the corresponding quotient space. In general (unless N is complemented[1] in E), the canonical projection $\Pi : E \mapsto E/N$ does not possess continuous linear right inverse operators $T : E/N \mapsto E$, that is, such that $\Pi T = I$. The following result (**Bartle & Graves theorem**) provides an often useful substitute for such a right inverse:

Theorem B.4 (Bartle and Graves, 1952; Zaidenberg et al., 1975) *For every $q > 1$ there exists a continuous homogeneous (but not necessarily linear[2]) map T :*

[1] That is, there exists a closed subspace M in E such that E is the direct sum $N \oplus N$. It is known that any Banach space not isomorphic to a Hilbert one, contains non-complemented subspaces.

[2] Additivity is missing.

$E/N \mapsto E$ such that $\Pi T = I$ and

$$\|T(x)\| \leq q\|x\| \text{ for all } x \in e/N$$

Corollary B.5 *For any surjective operator $A \in L(E, F)$ there exists a continuous homogeneous (not necessarily additive) bounded right inverse mapping $T : F \mapsto E$ (i.e., $AT = I$).*

Here is a stronger version of Theorem B.4:

Theorem B.6 *Let E and F be Frechet spaces and $A : E \mapsto F$ be a continuous linear surjective mapping. Then it possesses a homogeneous continuous right inverse mapping $T : F \mapsto E$.*

The following famous Banach open mapping theorem is also frequently needed.

Theorem B.7 *Every continuous surjective linear mapping of Frechet spaces is open.*

It can also be derived from a stronger version of Theorem B.4.

B.4 Fredholm and Semi-Fredholm Operators: Fredholm Index

Details of what is presented in this section can be found, for instance, in Gohberg and Kreĭn (1957); Kato (1995); Krein (1982); Zaidenberg et al. (1975).

Definition B.8 An operator $A \in L(E, F)$ is said to be **Fredholm** if

- $\dim \operatorname{Ker} A < \infty$;
- $\operatorname{Ran} A$ is closed;
- $\dim \operatorname{coKer} A < \infty$.

In this case

$$\operatorname{ind}(A) := \dim \operatorname{Ker} A - \dim \operatorname{coKer} A$$

is said to be the **index** of A.

The set of Fredholm operators acting from E into F will be denoted $\Phi(E, F) \subset L(E, F)$ ($\Phi(E) := \Phi(E, E)$). The set $\Phi^n(E, F)$ consists of Fredholm operators of index n. So, $\Phi = \bigcup_{n \in \mathbb{Z}} \Phi^n$. We also use the following notations for "**spectrum**" and "**co-spectrum**":

$$S_p\Phi(E) := \{A \in \Phi(E) \mid \dim \operatorname{Ker} A \geq p\} \tag{B.1}$$

and

$$CS_p\Phi(E) = \{A \in \Phi(E) \mid \dim \operatorname{coKer} A \geq p\} \tag{B.2}$$

The next result collects some standard properties of the Fredholm operators and their indexes:

Theorem B.9

(1) *Each of the sets* Φ^n *(and thus the whole* Φ*) is open in the operator norm topology.*
(2) *The index* ind *is a continuous (and thus locally constant) function on* $\Phi(E, F)$.
(3) $\text{ind}(AB) = \text{ind}(A) + \text{ind}(B)$.
(4) $A \in \Phi(E)$ *if and only if* $A^* \in \Phi(E^*)$*; furthermore,* $\text{ind}(A^*) = -\text{ind}(A)$.
(5) *If* $A \in \Phi(E, F)$ *and* $B \in C(E, F)$*, then* $(A + B) \in \Phi(E, F)$ *and* $\text{ind}(A + B) = \text{ind}(A)$.
(6) *The set* $\Phi^0(E, F)$ *consists of the sums* $A + B$*, where* $A \in L(E, F)$ *is invertible and* $B \in C(E, F)$*).*
(7) *An operator* $A \in L(E, F)$ *is Fredholm if and only if there exists an operator* $B \in L(F, E)$ *(called* **regularizer** *of* A*) such that* $AB - I$ *and* $BA - I$ *are compact operators.*
(8) *An operator* $A \in L(E, F)$ *is a Fredholm operator of index zero (i.e.,* $A \in \Phi^0(E, F)$*) if and only if it has an* **invertible regularizer**.

This theorem in particular claims that any Fredholm operator of zero index can be perturbed by addition of a compact (or even finite-dimensional) operator to become invertible. Proving it is a simple exercise.

The following facts are also often useful:

Lemma B.10

(1) *The functions* dim Ker A *and* dim coKer A *are upper semi-continuous on* $\Phi(E, F)$.
(2) *Let* $A_0 \in \Phi(E, F)$ *and* M *be a closed subspace of finite codimension in* E *such that* $M \cap \text{Ker}\, A = \{0\}$. *Then for all operators* A *in a neighborhood of* A_0 *one has* $M \cap \text{Ker}\, A = \{0\}$.

The second statement of the lemma is sometimes called Atiyah's lemma, due to its appearance in Atiyah (1989), although it most probably had been known before.

A wider class is formed by the **semi-Fredholm operators**.

Definition B.11 The class $\Phi_+(E, F)$ (correspondingly $\Phi_-(E, F)$) consists of all continuous linear operators from E to F with closed range and finite-dimensional kernel (correspondingly, co-kernel).

Under the additional condition that the range (correspondingly, the kernel) is a complemented subspace in F (in E), one obtains the class $\Phi_l(E, F)$ (respectively, $\Phi_r(E, F)$).

In the Hilbert case $\Phi_+(E, F) = \Phi_l(E, F)$ and $\Phi_-(E, F) = \Phi_r(E, F)$, due to existence of complements to all closed sub-spaces. Thus, in the Hilbert case, the classes Φ_{+-} are superfluous.

The names used are correspondingly **left semi-Fredholm** and **right semi-Fredholm** operators. The names chosen are explained by the following result:

Lemma B.12 *An operator $A \in L(E, F)$ belongs to $\Phi_l(E, F)$ if and only if there exists an operator $B \in L(F, E)$ (called* **left regularizer** *of A) such that the operator $BA - I$ is compact.*

Similarly, an operator $A \in L(E, F)$ belongs to $\Phi_r(E, F)$ if and only if there exists an operator $B \in L(F, E)$ (called **right regularizer** *of A) such that the operator $AB - I$ is compact.*

B.5 Schatten–von Neumann Classes and Regularized Determinants

We assume here that we are dealing with operators in a separable Hilbert space H. Let $A \in C(H)$ be a compact linear operator in H. For any such operator, we denote by $\lambda_1(A), \lambda_2(A), \ldots$ its eigenvalues counted with their multiplicity and arranged in the order of non-increasing magnitude. Then operator A^*A is a non-negative compact operator and let $(A^*A)^{1/2}$ be its non-negative square root.

Definition B.13 Let j be a non-negative integer. The j-th **singular value** (or the j-th **s-number**) of A is

$$s_j(A) = \lambda_j((A^*A)^{1/2}). \tag{B.3}$$

The importance of singular values is underscored by the following alternative descriptions:

Theorem B.14

(1)

$$s_j(A) = \min_{\dim M = j-1} \max_{x \neq 0,\, x \perp M} \frac{\|Ax\|}{\|x\|}.$$

(2)

$$s_j(A) = \min\{\|A - K\| \mid K \in L(H), \operatorname{rank} K \leq j - 1\}.$$

In other words, the singular values show the distances from operator A ***to operators of finite ranks.***

(3) *There exist (finite or infinite) orthonormal sequences of vectors $\{\phi_j\}$ and $\{\psi_j\}$ such that*

$$Ax = \sum_j s_j(A)(x \cdot \psi_j)\phi_j \qquad \text{for any } x \in H. \tag{B.4}$$

Remark B.15 Singular values have been used in many applications of numerical analysis for quite some time. The last statement of the previous theorem is called usually there **SVD (singular value decomposition)** and is one of the major tools in numerical linear algebra, as in many other areas (see, e.g., Strang, 1980).

We are now ready to introduce the celebrated Schatten–von Neumann classes of compact operators (Dunford and Schwartz, 1988; Gohberg and Kreĭn, 1969; Simon, 2005).

Definition B.16 Let $p \in [1, \infty]$. A compact linear operator A in H belongs to the $\mathcal{S}_p(H)$, if its the sequence of the singular values $s_j(A)$ belongs to l_p, that is,

$$\|A\|_p := \left(\sum_j s_j^p(A)\right)^{1/p} < \infty \text{ when } p < \infty.$$

For $p = \infty$, the norm naturally becomes $\|A\|_\infty := \max s_j(A)$, which coincides with the operator norm $\|A\|$ and thus $\mathcal{S}_\infty(H) = C(H)$.

The class $\mathcal{S}_p$ is a Banach space with respect to the norm $\|A\|_p$. When $p_1 < p_2$, the space $\mathcal{S}_{p_1}$ is naturally continuously embedded into $\mathcal{S}_{p_2}$.

The class $\mathcal{S}_1(H)$ consists of **trace class operators**, while $\mathcal{S}_2(H)$ consists of **Hilbert–Schmidt operators**.

The classes $\mathcal{S}_p$ are **operator ideals**, that is, if $A \in \mathcal{S}_p(H)$ and $B \in L(H)$, then $AB \in \mathcal{S}_p(H)$ and $BA \in \mathcal{S}_p(H)$.

One can introduce in a similar manner Schatten–von Neumann classes for compact operators acting between two Hilbert spaces.

The following example is instructive and very useful in this text. Consider the Sobolev space $H^s(\mathbb{T})$ of functions on the torus $\mathbb{T} = \mathbb{R}^n/\mathbb{Z}^n$. It consists of all functions $f(x) = \sum_{\gamma \in \mathbb{Z}^n} f_\gamma e^{2\pi i \gamma \cdot x}$ such that

$$\|f\|_s^2 = \sum (1 + |\gamma|)^{2s} |f_\gamma|^2 < \infty.$$

Theorem B.17 *Let $s > 0$. The natural embedding operator from $H^s(\mathbb{T})$ into $L_2(\mathbb{T})$ belongs to the class $\mathcal{S}_p(H^s(\mathbb{T}), L_2(\mathbb{T}))$ for $p > \frac{n}{s}$.*

The proof consists of a straightforward application of Fourier series expansions. A similar statement holds also on other compact manifolds, not just on the torus. The torus is chosen here, since it is our main case to consider, and besides, the proof in this case is very simple.

B.5.1 Regularized Determinants

Definition B.18 Let p be a positive integer and $A \in \mathcal{S}_p(H)$. We define the **regularized determinant** of the operator $I - A$ as follows:

$$\det_p(I - A) := \prod_j \left((1 - \lambda_j(A)) \exp\left(\sum_{k=1}^{p-1} k^{-1} \lambda_j^k(A) \right) \right). \tag{B.5}$$

If $A \in \mathcal{S}_r(H)$ for some non-integer r, the regularized determinant $\det_p(I - A)$ is defined for any $p > r$.

Remark B.19

- Readers familiar with Weierstrass' construction (see, e.g., Ch. 5, Th. 7 of Ahlfors, 1978) of an entire function with prescribed sequence of zeros (without accumulation points) would notice its similarity with the above definition of the regularized determinant. This is not a coincidence.
- The exponential terms in the formula correspond to the regularization, since without them the expression will diverge. However, if A is of the trace class, that is, $A \in \mathcal{S}_1$, the correction terms are not needed and the "usual" determinant is correctly defined as

$$\det_p(I - A) := \prod_j (1 - \lambda_j(A)).$$

- If $A_1, A_2 \in \mathcal{S}_p$ and $\sigma(A_1) = \sigma(A_2)$, then

$$\det_p(I - A_1) = \det_p(I - A_2). \tag{B.6}$$

The following result collects the main properties of regularized determinants (Dunford and Schwartz, 1988; Gohberg and Kreĭn, 1969; Simon, 2005):

Theorem B.20

(1) *If $r \leq p$, then the determinant $\det_p(I - A)$ is analytic as a complex-valued function on the complex Banach space $\mathcal{S}_p(H)$.*

(2) *If $p - 1 \leq r \leq p$, then*

$$|\det_p(I - A)| \leq C e^{C\|A\|_r^r},$$

where the constant C depends only on p and r.

(3) *If $p \geq 2, r \leq p - 1$, then for $A \in \mathcal{S}_r(H)$ one has*

$$\det_p(I - A) = e^{(p-1)^{-1} \mathrm{tr}(A^{p-1})} \det_{p-1}(I - A)$$

(4) *If $r \in [p-1, p]$ and $A \in \mathcal{S}_r(H)$, then the operator-valued function*

$$B_p(A) := \det_p(I - A)(I - A)^{-1} \tag{B.7}$$

(a priori defined only when $1 \notin \sigma(A)$), is analytic on the whole Banach space $\mathcal{S}_r(H)$ and satisfied the estimate

$$\|B_p(A)\| \leq e^{C(\|A\|_r^r + 1)}$$

for some constant C depending on p and r.

Corollary B.21 *Let $A \in \mathcal{S}_p(H)$ and $r \in [p-1, p]$, then*

(1)

$$\left|\det_p(I - kA)\right| \leq Ce^{C|k|^r \|A\|_r^r} \tag{B.8}$$

for any $k \in \mathbb{C}$ and an appropriate constant C.

(2) *The resolvent $(I - A)^{-1}$ (assuming that $1 \notin \sigma(A)$) can be represented as the ratio*

$$(I - A)^{-1} = B_p(A)/\det_p(A), \tag{B.9}$$

where both functions $B_p(A)$ and $\det_p(A)$ are analytic on $\mathcal{S}_p$ and satisfy the estimates

$$\|B_p(A)\|,\ |\det_p(A)| \leq e^{C\|A\|_p^p}. \tag{B.10}$$

(3) *Analogously,*

$$(I - kA)^{-1} = B_p(kA)/\det_p(kA), \tag{B.11}$$

where both functions $B_p(kA)$ and $\det_p(kA)$ are analytic on $\mathbb{C} \times \mathcal{S}_p$ and satisfy the estimates

$$\|B_p(kA)\|,\ |\det_p(kA)| \leq e^{C|k|^p \|A\|_p^p}. \tag{B.12}$$

We also need some multi-parameter resolvents of the form

$$L(k) := \left(1 - \sum_{j \in \mathbb{Z}_+^n |j| \leq m} k^j A_j\right)^{-1}, \tag{B.13}$$

where $k \in \mathbb{C}^n$ is a vector (quasimomentum in this book's setting), j is a multi-index with non-negative components, and $k^j = k_1^{j_1} \times \cdots \times k_n^{j_n}$. One can rewrite this operator-function as a **linear** operator-function in a finite sum of copies of the original space.

Theorem B.22 *In* (B.13), *let* $A_j = B_j S^{|j|}$, *where* $S \in \mathcal{S}_p$ *and* $B_j \in L(H)$. *Then*

$$L(k)^{-1} = B(k)/\delta(k), \tag{B.14}$$

where $B : \mathbb{C}^n \mapsto L(H)$ *and* $\delta : \mathbb{C}^n \mapsto \mathbb{C}$ *are entire functions, such that*

$$\|B(k)\|, |\delta(k)| \leq Ce^{C|\lambda|^p}. \tag{B.15}$$

Proof Let $\alpha := (\alpha_1, \ldots, \alpha_n) \in \mathbb{Z}_+^n$ and consider vectors

$$\psi := \{\{\psi_\alpha\} \mid |\alpha| \leq m - 1, \psi_\alpha \in \mathcal{H}\}. \tag{B.16}$$

These vectors form a finite direct sum $\mathcal{H}_1$ of several copies of $\mathcal{H}$.

For any $\alpha \neq (0, \ldots, 0)$, we denote by $j(\alpha)$ the first among the numbers j such that

$$\alpha_{j(\alpha)} = \max_i \alpha_i. \tag{B.17}$$

We also denote by α' the multi-index, which coincides with α, except the component $(\alpha)_{j(\alpha)}$, which is defined to be equal to $\alpha_{j(\alpha)} - 1$. In other words, the prime over the multiindex means decreasing by one its larger component.

Let us define $\psi_{(0,\ldots,0)} = u \in \mathcal{H}$. The equation $L(k)u = v$ can now be equivalently rewritten as a system for vectors in this space as follows:

$$\psi_0 + \sum_\alpha = v \tag{B.18}$$

$$\psi_\alpha - k_{j(\alpha)} S \psi_{\alpha'} = 0 \text{ for all } \alpha \neq (0, \ldots, 0). \tag{B.19}$$

The operator matrix on the left has the form

$$M(k) = I + \sum_j k_j \tilde{S}_j. \tag{B.20}$$

The entries of the operator-matrices $\tilde{S}_j$ are of the $\mathcal{S}_p$ class, and thus the matrices themselves also belong to $\mathcal{S}_p$. Now Theorem B.20 finishes the proof. □

Appendix C
Operator Functions

C.1 Operators Depending Smoothly on a Parameter

Theorem C.1 *Let T be a C^∞-manifold and $P : T \to L(H_1, H_2)$ be a C^∞-function with values in the space $L(H_1, H_2)$ of bounded linear operators between the Hilbert spaces H_1 and H_2. Assume that for each $k \in T$ the operator $P(k)$ is right semi-Fredholm (e.g.,* Zaidenberg et al., 1975*), that is, it has a closed range and a finite-dimensional cokernel. Then*

(1) *If $P(k)$ is surjective for all points k in T, then the multiplication operator*

$$C^\infty(T, H_1) \overset{P(k)}{\to} C^\infty(T, H_2) \tag{C.1}$$

is surjective.

(2) *For any fixed $k_0 \in T$ the dimension of the space of functionals of the form*

$$\left[\sum_{j \le N} D_{j,k}(\langle g_j, \phi \rangle) \right]_{k_0} \tag{C.2}$$

that are orthogonal to the range of the multiplication operator (C.1) *is finite. Here g_j are continuous linear functionals on H_2, the pairing $\langle g_j, \phi \rangle$ denotes the duality between H_2^* and H_2, $D_{j,k}$ are linear differential operators with respect to k on T, and $N \ge 0$.*

(3) *If $P(k)$ is surjective for all points k except of a finite subset $F \subset T$, then any continuous linear functional g on the space of smooth vector functions $C^\infty(T, H_2)$ that annihilates the range of the multiplication operator*

$$C^\infty(T, H_1) \overset{P(k)}{\to} C^\infty(T, H_2) \tag{C.3}$$

has the form

$$\langle g, \phi \rangle = \sum_{k_l \in F} \left[\sum_{j \leq N} D_{j,k}(\langle g_j, \phi \rangle) \right]_{k_l}, \tag{C.4}$$

in the notations of the previous statement of the lemma.

Proof Let us establish the validity of the first statement of the lemma. Due to the existence of partitions of unity, the statement is local. Locally, following for instance the proof of Theorem 2.7 in Zaidenberg et al. (1975), one can construct a smooth one-sided (right) inverse $Q(k)$ to the operator function $P(k)$. Now multiplication by $Q(k)$ provides a right inverse to (C.1), which proves the surjectivity.

To prove the second statement, let us consider a closed subspace $\mathrm{M} \subset H_1$ complementary to the kernel of $P(k_0)$. Then the operator $P(k_0) : \mathrm{M} \to H_2$ is injective and Fredholm. This injectivity property is preserved in a neighborhood U of k_0. In particular, the subspace $M(k) = P(k)(\mathrm{M}) \subset H_2$ of finite codimension forms a smooth subbundle $\mathcal{M}$ in $U \times H_2 \to U$. Now, any smooth H_2-valued function $f(k)$ such that $f(k) \in M(k)$, belongs to the range of the operator (C.1). Hence, functionals orthogonal to the range can be pushed down to the smooth sections of the finite-dimensional bundle $C = \bigcup_{k \in U} H_2/M(k)$ over U. It is clear that the functional one gets on this bundle preserves the structure (C.2). Such functionals on a finite-dimensional bundle, however, form a finite-dimensional space (for any fixed N).

The third statement can be proven analogously to the similar statement in Kuchment and Pinchover (2001). Namely, under the conditions of the statement, and taking into account the first claim of the lemma, any functional annihilating the range of the operator of multiplication by $P(k)$ must be supported at the finite set F over which $P(k)$ is not surjective. We can reduce the consideration to a neighborhood U of a single point $k_0 \in F$. Now, the proof of the second statement reduces the functional to one supported at the point k_0 and defined on smooth sections of a finite-dimensional bundle C over U. Thus, the standard representation of distributions supported at a point implies (C.4). □

C.2 Analytic Fredholm Operator Functions

All material (and more) of this section can be found in Zaidenberg et al. (1975).

Let $A(z)$, $z \in X$ be an analytic function on analytic manifold X with values in the set $\Phi(E)$ of Fredholm operators in a complex Banach space E. We are interested in the sets of points $z \in X$, where the kernel of $A(z)$ is nontrivial (the **spectrum of the operator function** $A(\cdot)$) and where the co-kernel is nontrivial (**co-spectrum**).

In order to do this, instead of working with particular operator functions, let us turn to the sets $S_p\Phi(E)$ and $CS_p\Phi(E)$ (see (B.1) and (B.2)).

Theorem C.2

(1) *The set* $S_p\Phi(E)$ *is an analytic subset of* $\Phi(E)$.
(2) *The same is true for* $CS_p\Phi(E)$.
(3) *In particular, the set* $\Phi^{(0)}(E)\setminus GL(E)$ *is a principal analytic subset in* $\Phi^{(0)}(E)$.

Let now $A(z), z \in X$ be a Fredholm-valued analytic function on analytic manifold X. We define scales of its **spectra (co-spectra)** as follows:

Definition C.3

- $$S_p(A(\cdot)) := \{z \in X \mid A(z) \in S_p\Phi(E)\} = A^{-1}(S_p\Phi(E)) = \{z \in X \mid \dim \operatorname{Ker} A(z) \geq p\}.$$
- $$CS_p(A(\cdot)) := \{z \in X \mid A(z) \in CS_p\Phi(E)\} = A^{-1}(CS_p\Phi(E)) = \{z \in X \mid \dim \operatorname{coKer} A(z) \geq p\}.$$

Corollary C.4

- $S_p(A(\cdot))$ *and* $CS_p(A(\cdot))$ *are analytic subsets of* X.
- *If the values* $A(z)$ *are operators of zero index, then* $S_1(A(\cdot)) = CS_1(A(\cdot))$ *is a principal analytic subset of* X.

Here is another often useful fact about analytic Fredholm operator functions:

Theorem C.5 *Let* Ω *be a Stein manifold and* $A(z), z \in \Omega$ *be an analytic operator function with values in the set of Fredholm operators in a Banach space* E. *If the minimum dimension of the kernel* Ker $A(z)$ *is not zero, then there exists an analytic vector-function* $F : \Omega \mapsto E$, *which is not identically equal to zero and such that* $A(z)F(z) = 0$ *for all* $z \in \Omega$.

Appendix D

Banach (Locally Trivial) Vector Bundles

This is a brief sketch of main notions and results concerning Banach locally trivial bundles. For more details, one can look into the survey Zaidenberg et al. (1975) or Kuchment (1993, Section 1.3).

We assume that the reader is familiar with the notion of a locally trivial vector bundle (see, e.g., Atiyah, 1989; Husemoller, 1975). The notion of a Banach bundle naturally generalizes it to the case of Banach (infinite-dimensional) fibers.

For all our purposes, it is sufficient to assume bases $\mathcal{X}$ of such bundles are connected, which we will silently do in the following.

Let $\mathcal{X}$ and $\mathcal{E}$ be topological spaces and $\pi : \mathcal{E} \mapsto \mathcal{X}$ be a continuous surjection, which we will call **projection**.

Definition D.1 We will call $\pi : \mathcal{E} \mapsto \mathcal{X}$ a **locally trivial (Banach) bundle** with a Banach space fiber E if a trivializing open cover $\{\mathcal{U}_j\}$ of $\mathcal{X}$ is given and for any j a local **trivializing map** $\phi_j : \pi^{-1}(\mathcal{U}_j) \mapsto \mathcal{U}_j \times E$, such that the following conditions are fulfilled:

- The following diagram is commutative:

$$\begin{array}{ccc} \pi^{-1}(\mathcal{U}_j) & \xrightarrow{\phi_j} & \mathcal{U}_j \times E \\ \downarrow{\scriptstyle \pi} & \swarrow{\scriptstyle \nu} & \\ \mathcal{U}_j & & \end{array} \tag{D.1}$$

  where ν is the natural projection of the product $\mathcal{U}_j \times E$ onto its the first factor.
- If $\mathcal{U}_i$ and $\mathcal{U}_j$ are two intersecting members of the cover and $\mathcal{U}_{ij}$ is their intersection, then the **transition function**

$$x \in \mathcal{U}_{ij} \mapsto \phi_{ij} := (\phi_j \phi_i^{-1}|_{\pi^{-1}(x)}) \in L(E)$$

is continuous with respect to the norm topology in $L(E)$

For any $x \in \mathcal{X}$, the **fiber** $\pi^{-1}(x)$ can be identified by the trivialisation map with the Banach space E.

The simplest example is the **trivial bundle** $\mathcal{X}\times E \mapsto \mathcal{X}$ with the natural projection on the first factor.

D.1 Morphisms

Definition D.2

- A mapping $A : \mathcal{E}_1 \mapsto \mathcal{E}_2$ of two Banach bundles $\pi_1 : \mathcal{E}_1 \mapsto \mathcal{X}$ and $\pi_2 : \mathcal{E}_2 \mapsto \mathcal{X}$ over the same base $\mathcal{X}$ is a **morphism**, if
  (1) it preserves fibers, that is, $\pi_2 A = \pi_1$;
  (2) at each point $x \in \mathcal{X}$, the mapping A is a bounded linear operator $A(x)$ between the corresponding fibers;
  (3) if $\phi_j, j = 1, 2$ is a trivialization of $\mathcal{E}_j$in a neighborhood U of x, then the mapping $y \in U \mapsto \phi_2 \circ A(y) \circ \phi_1^{-1} \in L(E)$ is continuous (with respect the norm operator topology).
- An **isomorphism** is a morphism which is invertible on each fiber. If such an isomorphism exists, the **bundles are isomorphic**.
- A bundle isomorphic to $\mathcal{X} \times E \mapsto \mathcal{X}$ is called **trivializable** (or just **trivial**).
- A **morphism is Fredholm**, if it acts as a Fredholm operator on each fiber.

Example D.3 An important example of a trivial bundle is the standard projection map of a Banach space E onto its quotient over a closed subspace N:

$$E \mapsto E/N.$$

The validity of this example is not completely trivial. Indeed, if N is complemented in E, so $E = M \oplus N$, and thus M is isomorphic to E/N, we immediately get the definition of a trivial bundle with the base M: $M \oplus N \mapsto M \approx E/N$. However, if E is not isomorphic to a Hilbert space, N might have no complement. Despite this, the Bartle–Graves theorem B.4 allows one to still consider the projection as trivial Banach bundle with fibers isomorphic to N. We leave it to the reader to reconstruct the details.

Let us have an **infinite-dimensional** Banach (or Hilbert) bundle with fiber H. It is a simple exercise to prove that if the general linear group $G(H)$ is contractible in the operator norm topology, any such bundle is trivial (see, e.g., Atiyah, 1989,

Zaidenberg et al., 1975). The famous theorem by Kuiper (1965) (see also its discussion in Atiyah, 1989) claims that the general linear group $GL(H)$ of an infinite-dimensional Hilbert space is contractible, and so all such bundles are topologically trivial. Although it is not true for the general linear group $GL(B)$ of an arbitrary infinite-dimensional Banach space, thus nontrivial bundles with such fibers can exist, the contractibility is true for all (at least a majority of) "classical" infinite-dimensional Banach spaces (see the survey in Mitjagin, 1970). Luckily, we will be dealing with the Hilbert case solely.

D.2 Smooth and Holomorphic Bundles

Imposing on the transition functions smoothness or analyticity conditions, one arrives to the notions of smooth and analytic vector/Banach bundles. Correspondingly, one introduces the notions of topologically, or smoothly, or analytically trivial bundles. A bundle can be topologically, but not necessarily smoothly or analytically trivial. The following incarnation of the Oka's principle is crucial for us:

Theorem D.4 *An analytic Banach bundle over a Stein base is topologically trivial, then it is also analytically trivial. (The converse claim is obvious.)*

This famous theorem in the case of finite-dimensional bundles is due to Grauert (1958), and its infinite-dimensional version due to Bungart (1967).

Appendix E
The Landis Conjecture

The so-called Landis conjecture (see Kondrat′ev and Landis, 1988; Landis, 1963a, 1963b) asks the question of the possible rate of decay at infinity of solutions of equation,

$$-\Delta u + Vu = 0, \tag{E.1}$$

with potential $V \in L^\infty(\mathbb{R}^n)$. In particular, the **strong Landis conjecture** asks whether there is a constant C (depending on $\|V\|_{L^\infty}$)) such that any solution decaying as $|u(x)| \le e^{-D|x|}$ for some $D > C$, must vanish identically.

For this text, we are only interested in the **weaker** form of the Landis conjecture: If a solution u satisfies the estimate

$$|u(x)| \le e^{-|x|^\gamma} \tag{E.2}$$

for some $\gamma > 1$, then the solution must vanish identically.

Similar conjectures can be formulated for more general elliptic equations of the second order, not just the Schrödinger equation (E.1).

Why do we insist on equations of second order? The reason is that for higher-order elliptic equations, due to absence of uniqueness of continuation property, fast decaying and even compactly supported solutions can exist; see Pliś (1960) and discussions of the periodic case in Chapter 5 and Kuchment (1993, 2016).

In spite of many current efforts, this conjecture is still unproven. One can find very recent surveys (eternally incomplete, due to the avalanche of new papers) and new results in Fernandez–Bertolin et al. (2024); Krymskii et al. (2025); Logunov and Malinnikova (2020, 2018).

An important early breakthrough was achieved in Froese et al. (1983); Meshkov (1991), where it was proven for (E.2) that solutions decaying at the rate $e^{-|x|^\gamma}$ with $\gamma > 4/3$ must be identically equal to zero. Moreover, it was shown in Meshkov (1991) that the magical $4/3$ cannot be improved, albeit the provided example dealt with a **complex** potential. It had been rather common opinion that in self-adjoint case, any exponential order $\gamma > 1$ should be prohibited. This expectation was somewhat shaken by the result of Krymskii and Filonov (2023), where the rate $4/3$ was achieved in self-adjoint situation, albeit in a **half**-cylinder (see also Theorem 1.1 in Krymskii et al., 2025). Another warning sign was given by Theorem 1.2 of Krymskii et al. (2025), where an example of **double-exponential**(!) decay of eigenfunction was constructed, if one enforces periodicity in all but one dimensions. This shows that if the weaker Landis conjecture conjecture does holds in our periodic situations, periodicity in all variables should play a major role.

Appendix F

Proofs of Some Technical Statements

(Sketch of the) Proof of Theorem 4.4.4

(1) *There exists an entire function $f(k, \lambda)$ on $\mathbb{C}^{n+1}$, such that*

(a)

$$|f(k, \lambda)| \le C_p e^{(|k|+|\lambda|)^p} \tag{F.1}$$

for any $p > n$. (A similar statement holds for more general elliptic periodic operators, with the exponent p depending on the dimension and order of the operator [Kuchment, 1993].*)*

(b)

$$B_{H,\mathbb{C}} = \{(k, \lambda) \in \mathbb{C}^{n+1} \mid f(k, \lambda) = 0\}. \tag{F.2}$$

(2) *If $p > n + 1$, one can ensure that f is Γ^*-periodic with respect to the quasi-momentum k.*

In particular, $B_{H,\mathbb{C}}$ is a ***Γ^*-periodic complex analytic subvariety of $\mathbb{C}^{n+1}$ of co-dimension*** 1.

Sketch of the Proof First of all, let us fix for simplicity the value of the energy, say $\lambda = 0$. This reduces us to the case of the Floquet surface F_λ and prove existence of an entire function $f(k)$ with the properties declared in the theorem. This would easily generalize to include as an additional variable the spectral parameter, and thus establish the statement of the theorem.

The idea is to find function f as some kind of a regularized determinant (see Appendix B). So,

$$F_\lambda = \{k \in \mathbb{C}^n \mid \operatorname{Ker} H(x, D + k) \neq 0\} \neq \mathbb{C}^n, \tag{F.3}$$

where the kernel of the operator is considered in the Sobolev space $H^{2m}(\mathbb{T})$ (where

$2m$ is the order of the operator). Then, the operator polynomial can be written as

$$H(k) = H_0 + \sum_{|j|>0} k^j H_j,$$

where $H_0 = H(x, D)$ and H_j has the order $2m - |j|$. Suppose that H_0 is invertible (which can be achieved by a shift with respect to k, unless $F_\lambda = \mathbb{C}^n$, in which case the statement is trivial). Then

$$H(k) = H_0(I + \sum_{|j|>0} k^j H_0^{-1} H_j). \tag{F.4}$$

The order of $B_j := H_0^{-1} H_j$ is equal to $-|j|$. Let also S be an invertible pseudo-differential operator of order -1 on $\mathbb{T}$. Then

$$H(k) = H_0(I + \sum_{|j|>0} k^j B_j S^{|j|}), \tag{F.5}$$

where the operators $B_j = H_0^{-1} H_j S^{|j|}$ are of order zero, and thus bounded in any Sobolev space on the torus. Now Theorem B.22 finishes the proof. □

Sketch of the Proof of Theorem 5.5.1

Let $B_H = \bigcup_l B_l$ be the (locally finite) decomposition of the Bloch variety into irreducible components $B_l \subset \mathbb{C}^{n+1}$ (see Appendix A). Then, for each l, there exists a proper analytic subset A_l of B_l of dimension $(n-1)$, and a finite set of functions

$$\psi_{j,l,(k,\lambda)}(x), \quad j = 1, \ldots, j_m \tag{F.6}$$

such that:

(1) *Functions $\psi_{j,l,(k,\lambda)}(x)$ are analytic on $B_l \setminus A_l$ as functions of (k, λ) with values in the Sobolev space $H^s(\mathbb{T})$.*
(2) *For each $(k, \lambda) \in B_l \setminus A_l$, $\psi_{j,l,(k,\lambda)}(x) \in K_{k,\lambda}$.*
(3) *For each $(k, \lambda) \in B_l \setminus A_l$, the functions $\psi_{j,l,(k,\lambda)}(x)$, $j = 1, \ldots, j_m$ form a basis in $K_{k,\lambda}$.*

In particular, the fibration

$$\coprod_{(k,\lambda)\in B_l \setminus A_l} K_{k,\lambda} \mapsto (B_l \setminus A_l) \tag{F.7}$$

is a trivializable holomorphic vector subbundle of the trivial bundle over $(B_l \setminus A_l)$ with the fiber $H^s(\mathbb{T})$.

Sketch of the proof It is sufficient to consider one of the irreducible components $B := B_l$, which is automatically Stein. According to Kuchment (1993, Theorem 1.6.14), the sheaf of the germs of analytic sections of Ker $(H(k) - \lambda I)$ over B is coherent. Let the minimal dimension of this kernel over B be equal to k. Then, due to irreducibility of B, the dimension is the same outside of an analytic subset S of B. In a neighborhood of any point $(k_0, \lambda_0) \in B \setminus S$ there is a local holomorphic basis of sections $\phi_J(\lambda, k), j = 1, \dots, k$. Then, according to classical approximation theorem (see, e.g., Theorem 7.2.7 in Hörmander, 1990). These functions can be in a (maybe smaller) neighborhood of (k_0, λ_0) uniformly approximated by global sections ψ_j. Then, aside from a proper analytic subset A of B, they stay the desired basis sections. □

(Sketch of the) Proof of Theorem 5.7.1

(1) *The singular continuous spectrum of the operator A is empty*

$$\sigma_{sc}(A) = \emptyset. \tag{F.8}$$

(2) *The pure point spectrum consists of all values λ such that the λ-level set of $a(k)$ has positive measure. In particular, if there is no positive measure "flat piece" $a = const$, then $\sigma_{pp}(A) = \emptyset$ and thus the whole spectrum is absolutely continuous:*

$$\sigma(A) = \sigma_{ac}(A).$$

(3) *For each $\lambda \in \sigma_{pp}(A)$, the function $a(k)$ is constant on an open set.*

Proof (1) Absence of the singular continuous spectrum follows, for instance, from the direct calculation of the Radon–Nikodym derivative, as in Theorem 5.9.5.

(2) Existence of an eigenfunction clearly implies that the "symbol" $a(\lambda)$ must have constant value (the supposed eigenvalue) on a set of a positive measure. Analyticity implies existence of a flat piece. □

References

Adachi, T., and Sunada, T. 1993. Density of states in spectral geometry. *Comment. Math. Helv.*, **68**(3), 480–493.

Adams, S. D. M., Craster, R. V., and Guenneau, S. 2008. Bloch waves in periodic multi-layered acoustic waveguides. *Proc. R. Soc. Lond. Ser. A Math. Phys. Eng. Sci.*, **464**(2098), 2669–2692.

Agmon, S. 1984. On positive solutions of elliptic equations with periodic coefficients in $\mathbf{R}^n$, spectral results and extensions to elliptic operators on Riemannian manifolds. Pages 7–17 of: *Differential Equations (Birmingham, Ala., 1983)*. North-Holland Mathematics Studies, vol. 92. North-Holland.

Agmon, S. 1992. On the representation theorem for solutions of the Helmholtz equation on the hyperbolic space. Pages 1–20 of: *Partial Differential Equations and Related Subjects (Trento, 1990)*. Pitman Research Notes in Mathematics Series, vol. 269. Longman Scientific & Technical Publisher.

Agmon, S. 1999. Representation theorems for solutions of the Helmholtz equation on $\mathbb{R}^n$. Pages 27–43 of: *Differential Operators and Spectral Theory*. American Mathematical Society Translations: Series 2, vol. 189. American Mathematical Society.

Agmon, S. 2010. *Lectures on Elliptic Boundary Value Problems*. AMS Chelsea Publishing.

Agranovskiĭ, M. L. 1993. *Invariant Function spaces on Homogeneous Manifolds of Lie Groups and Applications*. Translated from the Russian manuscript by A. I. Zaslavsky. Translations of Mathematical Monographs, vol. 126. American Mathematical Society.

Ahlfors, L. V. 1978. *Complex Analysis*. 3rd ed. An Introduction to the Theory of Analytic Functions of One Complex Variable. International Series in Pure and Applied Mathematics. McGraw-Hill Book Co.

Aizenman, M., and Warzel, S. 2015. *Random Operators*. Disorder Effects on Quantum Spectra and Dynamics. Graduate Studies in Mathematics, vol. 168. American Mathematical Society.

Akhiezer, N. I., and Glazman, I. M. 1993. *Theory of Linear Operators in Hilbert Apace*. Translated from the Russian and with a preface by Merlynd Nestell. Reprint of the 1961 and 1963 translations, two volumes bound as one. Dover Publications Inc.

Alama, S., Avellaneda, M., Deift, P. A., and Hempel, R. 1994. On the existence of eigenvalues of a divergence-form operator $A + \lambda B$ in a gap of $\sigma(A)$. *Asymptot. Anal.*, **8**(4), 311–344.

Alama, S., Deift, P. A., and Hempel, R. 1989. Eigenvalue branches of the Schrödinger operator $H - \lambda W$ in a gap of $\sigma(H)$. *Commun. Math. Phys.*, **121**(2), 291–321.

Alexander, H. 1975. On a problem of Julia. *Duke Math. J.*, **42**, 327–332.

Allaire, G., Conca, C., and Vanninathan, M. 1998. The Bloch transform and applications. Pages 65–84 of: *Actes du 29ème Congrès d'Analyse Numérique: CANum'97 (Larnas, 1997)*. ESAIM Proceedings, vol. 3. Society for Applied and Industrial Mathematics.

Arlinskii, Y., Belyi, S., and Tsekanovskii, E. 2011. *Conservative Realizations of Herglotz–Nevanlinna functions*. Operator Theory: Advances and Applications, vol. 217. Birkhäuser/Springer Basel AG.

Arnol′d, V. I. 1978. *Ordinary Differential Equations*. Translated from the Russian and edited by Richard A. Silverman. MIT Press.

Arnol′d, V. I. 1988. *Geometrical Methods in the Theory of Ordinary Differential Equations*. 2nd ed. Translated from the Russian by Joseph Szücs [József M. Szücs]. Grundlehren der Mathematischen Wissenschaften, vol. 250. Springer-Verlag.

Arnol′d, V. I. 1998. On the teaching of mathematics. *Uspekhi Mat. Nauk.*, **53**(1(319)), 229–234.

Arscott, F. M. 1964. *Periodic Differential Equations: An Introduction to Mathieu, Lamé, and Allied Functions*. International Series of Monographs in Pure and Applied Mathematics, vol. 66. Pergamon Press.

Asch, J., and Knauf, A. 1998. Motion in periodic potentials. *Nonlinearity*, **11**(1), 175–200.

Ashcroft, N. W., and Mermin, N. D. 1976. *Solid State Physics*. Holt, Rinehart and Winston.

Atiyah, M. F. 1989. *K-Theory*. 2nd ed. Notes by D. W. Anderson. Addison-Wesley Publishing Company.

Atiyah, M. F., and Segal, G. B. 1968. The index of elliptic operators. II. *Ann. Math. (2)*, **87**, 531–545.

Atiyah, M. F., and Singer, I. M. 1968a. The index of elliptic operators. I. *Ann. Math. (2)*, **87**, 484–530.

Atiyah, M. F., and Singer, I. M. 1968b. The index of elliptic operators. III. *Ann. Math. (2)*, **87**, 546–604.

Auckly, D., and Kuchment, P. 2018. On Parseval frames of exponentially decaying composite Wannier functions. Pages 227–240 of: *Mathematical Problems in Quantum Physics*. Contemporary Mathematics, vol. 717. American Mathematical Society

Avron, J. E., and Simon, B. 1978. Analytic properties of band functions. *Ann. Phys.*, **110**(1), 85–101.

Bakhvalov, N., and Panasenko, G. 1989. *Homogenization: Averaging Processes in Periodic Media. Mathematical Problems in the Mechanics of Composite Materials*. Translated from the Russian by D. Leĭtes. Mathematics and Its Applications (Soviet Series), vol. 36. Kluwer Academic Publishers Group.

Ballas, S. A., Needham, T., and Schonkwiller, C. 2023. On the existence of Parseval frames for vector bundles. ArXiv:2312.13488.

Barbatis, G., and Parnovski, L. 2009. Bethe–Sommerfeld conjecture for pseudodifferential perturbation. *Comm. Partial Differ. Equ.*, **34**(4–6), 383–418.

Bartle, R. G., and Graves, L. M. 1952. Mappings between function spaces. *Trans. Amer. Math. Soc.*, **72**, 400–413.

Barut, A. O., and Râczka, R. 1986. *Theory of Group Representations and Applications*. 2nd ed. World Scientific Publishing Co.

Bassani, F., and Pastori Parravicini, G. 1975. *Electronic States and Optical Transitions in Solids*. Pergamon Press.

Bättig, D., Knörrer, H., and Trubowitz, E. 1991. A directional compactification of the complex Fermi surface. *Compos. Math.*, **79**(2), 205–229.

Behrndt, J., Schmitz, P., and Teschl, G., and Trunk, C. 2023. Perturbations of periodic Sturm–Liouville operators. *Adv. Math.*, **422**, Paper No. 109022, 22.

Bensoussan, A., Lions, J.-L., and Papanicolaou, G. 1978. *Asymptotic Analysis for Periodic Structures*. Studies in Mathematics and its Applications, vol. 5. North-Holland Publishing Co.

Bentosela, F., Duclos, P., and Exner, P. 2003. Absolute continuity in periodic thin tubes and strongly coupled leaky wires. *Lett. Math. Phys.*, **65**(1), 75–82.

Berezanskii, Ju. M. 1968. *Expansions in Eigenfunctions of Self-Adjoint Operators*. Translations of Mathematical Monographs, vol. 17. American Mathematical Society.

Berezansky, Y. M., Sheftel, Z. G., and Us, G. F. 1996. *Functional Analysis. Vol. II*. Translated from the 1990 Russian original by Peter V. Malyshev. Operator Theory: Advances and Applications, vol. 86. Birkhäuser Verlag.

Berezin, F. A., and Shubin, M. A. 1991. *The Schrödinger Equation*. Translated from the 1983 Russian edition by Yu. Rajabov, D. A. Leĭtes and N. A. Sakharova and revised by Shubin. With contributions by G. L. Litvinov and Leĭtes. Mathematics and Its Applications (Soviet Series), vol. 66. Kluwer Academic Publishers Group.

Berkolaiko, G., and Comech, A. 2018. Symmetry and Dirac points in graphene spectrum. *J. Spectr. Theory*, **8**(3), 1099–1147.

Berkolaiko, G., and Kuchment, P. 2013. *Introduction to Quantum Graphs*. Mathematical Surveys and Monographs, vol. 186. American Mathematical Society.

Berkolaiko, G., and Zelenko, I. 2024. Morse inequalities for ordered eigenvalues of generic self-adjoint families. *Invent. Math.*, **238**(1), 283–330.

Berkolaiko, G., Carlson, R., Fulling, S., and Kuchment, P. (eds). 2006. *Quantum Graphs and Their Applications*. Contemporary Mathematics, vol. 415. American Mathematical Society.

Bernevig, B. A. 2013. *Topological Insulators and Topological Superconductors*. Princeton University Press.

Berry, M. V. 1990. Anticipations of the geometric phase. *Physics Today*, December, 34–40.

Berry, M. V., and Wilkinson, M. 1984. Diabolical points in the spectra of triangles. *Proc. R. Soc. Lond. A Math. Phys. Sci.*, **392**(1802), 15–43.

Bers, L., John, F., and Schechter, M. 1979. *Partial Differential Equations*. With supplements by Lars Gårding and A. N. Milgram, with a preface by A. S. Householder, Reprint of the 1964 original. Lectures in Applied Mathematics, vol. 3. American Mathematical Society.

Bethe, H., and Sommerfeld, A. 1933. Elektronentheorie der Metalle. Pages 333–622 of: *Handbuch der Physik*, vol. 24,2, Geiger, H., and Scheel, K. (eds), Springer. Reprinted as *Elektronentheorie der Metalle*, Springer, 1967.

Bierstone, E., and Milman, P. D. 1988. Semianalytic and subanalytic sets. *Inst. Hautes Études Sci. Publ. Math.*, **67**, 5–42.

Bierstone, E., and Milman, P. D. 1991. A simple constructive proof of canonical resolution of singularities. Pages 11–30 of: *Effective Methods in Algebraic Geometry (Castiglioncello, 1990)*. Progress in Mathematics, vol. 94. Birkhäuser Boston.

Birman, M. Sh. 1991a. Discrete spectrum in a gap of perturbed periodic operator at large coupling constants. Pages 16–24 of: *Rigorous Results in Quantum Dynamics (Liblice, 1990)*. World Scientific Publishing.

Birman, M. Sh. 1991b. On a discrete spectrum in gaps of a second-order perturbed periodic operator. *Funktsional. Anal. i Prilozhen.*, **25**(2), 89–92.

Birman, M. Sh. 1994. Discrete spectrum of the periodic elliptic operator with a differential perturbation. Pages Exp. No. XIV, 4 of: *Journées "Équations aux Dérivées Partielles" (Saint-Jean-de-Monts, 1994)*. École Polytech., Palaiseau.

Birman, M. S. 1995. The discrete spectrum in gaps of the perturbed periodic Schrödinger operator. I. Regular perturbations. Pages 334–352 of: *Boundary Value Problems, Schrödinger Operators, Deformation Quantization*. Math. Top., vol. 8. Akademie Verlag.

Birman, M. S. 1996. The discrete spectrum of the periodic Schrödinger operator perturbed by a decreasing potential. *Algebra i Analiz*, **8**(1), 3–20.

Birman, M. S. 1997. The discrete spectrum in gaps of the perturbed periodic Schrödinger operator. II. Nonregular perturbations. *Algebra i Analiz*, **9**(6), 62–89.

Birman, M. S., and Raĭkov, G. D. 1991. Discrete spectrum in the gaps for perturbations of the magnetic Schrödinger operator. Pages 75–84 of: *Estimates and Asymptotics for Discrete Spectra of Integral and Differential Equations (Leningrad, 1989–90)*. Advances in Soviet Mathematics, vol. 7. American Mathematical Society.

Birman, M. S., and Sloushch, V. A. 2010. Discrete spectrum of the periodic Schrödinger operator with a variable metric perturbed by a nonnegative potential. *Math. Model. Nat. Phenom.*, **5**(4), 32–53.

Birman, M. S., and Suslina, T. 1999a. A periodic magnetic Hamiltonian with a variable metric. The problem of absolute continuity. *Algebra i Analiz*, **11**(2), 1–40. Translation in *St. Petersburg Math. J.*, **11**(2), (2000), 203–232.

Birman, M. S., and Suslina, T. A. 1999b. The periodic Dirac operator is absolutely continuous. *Integr. Equ. Oper. Theory*, **34**(4), 377–395.

Birman, M. S., and Suslina, T. A. 1999c. Two-dimensional periodic Pauli operator. The effective masses at the lower edge of the spectrum. Pages 13–31 of: *Mathematical Results in Quantum Mechanics (Prague, 1998)*. Operator Theory: Advances and Application, vol. 108. Birkhäuser.

Birman, M. S., and Suslina, T. A. 2000. On the absolute continuity of the periodic Schrödinger and Dirac operators with magnetic potential. Pages 41–49 of: *Differential Equations and Mathematical Physics (Birmingham, AL, 1999)*. AMS/IP Studies in Advanced Mathematics, vol. 16. American Mathematical Society.

Birman, M. S., and Suslina, T. A. 2002. Absolute continuity of the spectrum of the periodic operator of elasticity theory for constant shear modulus. Pages 69–74 of: *Nonlinear Problems in Mathematical Physics and Related Topics, II*. International Mathematics Series (N.Y.), vol. 2. Kluwer/Plenum.

Birman, M. S., and Weidl, T. 1994. The discrete spectrum in a gap of the continuous one for compact supported perturbations. Pages 9–12 of: *Mathematical Results in Quantum Mechanics (Blossin, 1993)*. Operator Theoretical Methods and Applications, vol. 70. Birkhäuser.

Birman, M. S., Suslina, T., and Shterenberg, R. 2000a. Absolute continuity of the two-dimensional Schrödinger operator with delta potential concentrated on a periodic system of curves. *Algebra i Analiz*, **12**(6), 140–177.

Birman, M. S., Laptev, A., and Suslina, T. A. 2000b. The discrete spectrum of a two-dimensional second-order periodic elliptic operator perturbed by a decreasing potential. I. A semi-infinite gap. *Algebra i Analiz*, **12**(4), 36–78.

Boas, Jr., R. P. 1954. *Entire Functions*. Academic Press Inc.

Borg, G. 1946. Eine Umkehrung der Sturm–Liouvillischen Eigenwertaufgabe. *Acta Math.*, **78**, 1–96.

Borisov, D. I., and Pankrashkin, K. V. 2013a. Quantum waveguides with small periodic perturbations: gaps and edges of Brillouin zones. *J. Phys. A*, **46**(23), 235203, 18.

Borisov, D. I., and Pankrashkin, K. V. 2013b. On the extrema of band functions in periodic waveguides. *Funktsional. Anal. i Prilozhen.*, **47**(3), 87–90.

Bouckaert, L. P., Smoluchowski, R., and Wigner, E. 1936. Theory of Brillouin zones and symmetry properties of wave functions in crystals. *Phys. Rev.*, **50**(1), 58–67.

Bourgain, J. 2005. *Green's Function Estimates for Lattice Schrödinger Operators and Applications*. Annals of Mathematics Studies, vol. 158. Princeton University Press.

Bourne, C., and Mesland, B. 2021. Localised module frames and Wannier bases from groupoid Morita equivalences. *J. Fourier Anal. Appl.*, **27**(4), Paper No. 69, 39.

Boyarchenko, S. I., and Levendorskiĭ, S. Z. 1997a. An asymptotic formula for the number of eigenvalue branches of a divergence form operator $A + \lambda B$ in a spectral gap of A. *Comm. Partial Differ. Equ.*, **22**(11–12), 1771–1786.

Boyarchenko, S. I., and Levendorskiĭ, S. Z. 1997b. Precise spectral asymptotics for perturbed magnetic Schrödinger operator. *J. Math. Pures Appl. (9)*, **76**(3), 211–236.

Brillouin, L. 1953. *Wave Propagation in Periodic Structures. Electric Filters and Crystal Lattices*. 2nd ed. Dover Publications Inc.

Brillouin, L., and Parodi, M. 1956. *Propagation des Ondes dans les Milieux Périodiques*. Masson et Cie,

Brillouin, L. 1960. *Wave Propagation and Group Velocity*. Pure and Applied Physics, vol. 8. Academic Press.

Brooks, R. 1981. The fundamental group and the spectrum of the Laplacian. *Comment. Math. Helv.*, **56**(4), 581–598.

Brooks, R. 1986. The spectral geometry of a tower of coverings. *J. Differential Geom.*, **23**(1), 97–107.

Brouder, C., Panati, G., Calandra, M., Mourougane, C., and Marzari, N. 2007. Exponential localization of Wannier functions in insulators. *Phys. Rev. Lett.*, **98**(4), 046402.

Brown, B. M., Eastham, M. S. P., and Schmidt, K. M. 2013. *Periodic Differential Operators*. Operator Theory: Advances and Applications, vol. 230. Birkhäuser/Springer.

Brüning, J., and Sunada, T. 1992a. On the spectrum of gauge-periodic elliptic operators. In *Méthodes Semi-Classiques, Vol. 2 (Nantes, 1991). Astérisque*, **210**, 6, 65–74.

Brüning, J., and Sunada, T. 1992b. On the spectrum of periodic elliptic operators. *Nagoya Math. J.*, **126**, 159–171.

Bungart, L. 1967. On analytic fiber bundles. I. Holomorphic fiber bundles with infinite-dimensional fibers. *Topology*, **7**, 55–68.

Busch, K., Mingaleev, S., Garcia-Martin, A., Schillinger, M., and Hermann, D. 2003. The Wannier function approach to photonic crystal circuits. *Journal of Physics: Condensed Matter*, **15**(30), R1233.

Callaway, J. 1964. *Energy Band Theory*. Pure and Applied Physics, vol. 16. Academic Press.

Carmona, R., and Lacroix, J. 1990. *Spectral Theory of Random Schrödinger Operators*. Probability and its Applications. Birkhäuser Boston, Inc.

Chirka, E. M. 1989. *Complex Analytic Sets*. Translated from the Russian by R. A. M. Hoksbergen. Mathematics and its Applications (Soviet Series), vol. 46. Kluwer Academic Publishers Group.

Chow, Shui-Nee, Lu, K., and Mallet-Paret, J. 1994. Floquet theory for parabolic differential equations. *J. Differential Equations*, **109**(1), 147–200.

Chow, Shui-Nee, Lu, K., and Mallet-Paret, J. 1995. Floquet bundles for scalar parabolic equations. *Arch. Rational Mech. Anal.*, **129**(3), 245–304.

Choy, T.-S., Naset, J., Chen, J., Hershfield, S. and Stanton, C. 2000. A database of Fermi surfaces in virtual reality modeling language (VRML). {em Bull. Amer. Phys. Soc., **45**(1), L36–42. `www.phys.ufl.edu/fermisurface/`.

Chulaevsky, V. A. 1989. *Almost Periodic Operators and Related Nonlinear Integrable Systems*. With a foreword by Ya. G. Sinaĭ, Translated from the Russian. Nonlinear Science: Theory and Applications. Manchester University Press, Manchester.

Coddington, E. A., and Carlson, R. 1997. *Linear Ordinary Differential Equations*. Society for Industrial and Applied Mathematics.

Coddington, E. A., and Levinson, N. 1955. *Theory of Ordinary Differential Equations*. McGraw-Hill Book Company, Inc.,

Colin de Verdière, Y. 1991. Sur les singularités de van Hove génériques. In *Analyse globale et physique mathématique (Lyon, 1989). Mém. Soc. Math. France (N.S.)*, **46**, 99–110.

Colin de Verdière, Y., and Kappeler, T. 1989. On double eigenvalues of Hill's operator. *J. Funct. Anal.*, **86**(1), 127–135.

Conca, C., Planchard, J., and Vanninathan, M. 1995. *Fluids and Periodic Structures*. RAM: Research in Applied Mathematics, vol. 38. John Wiley & Sons, Ltd.

Cornean, H. D., Herbst, I., and Nenciu, G. 2016. On the construction of composite Wannier functions. *Ann. Henri Poincaré*, **17**(12), 3361–3398.

Cornean, H. D., Monaco, D., and Teufel, S. 2017. Wannier functions and Z2 invariants in time reversal symmetric topological insulators. *Rev. Math. Phys.*, **29**, 1730001.

Cornean, H. D., Gontier, D., Levitt, A., and Monaco, D. 2019a. Localised Wannier functions in metallic systems. *Ann. Henri Poincare*, **20**(4), 1367–1391.

Cornean, H. D., Monaco, D., and Moscolari, M. 2019b. Parseval frames of exponentially localized magnetic Wannier functions. *Commun. Math. Phys.*, **371**, 1179–1230.

Corwin, L. J., and Greenleaf, F. P. 1990. *Representations of Nilpotent Lie Groups and their Applications. Part I: Basic Theory and Examples*. Cambridge Studies in Advanced Mathematics, vol. 18. Cambridge University Press.

Cracknell, A. P. 1971. *The Fermi Surfaces of Metals*. Taylor & Francis.

Cracknell, A. P., and Wong, K. S. 1973. *The Fermi Surface*. Clarendon Press.

Cycon, H. L., Froese, R. G., Kirsch, W., and Simon, B. 1987. *Schrödinger Operators with Application to Quantum Mechanics and Global Geometry*. Study edn. Texts and Monographs in Physics. Springer-Verlag.

Dahlberg, B. E. J., and Trubowitz, E. 1982. A remark on two-dimensional periodic potentials. *Comment. Math. Helv.*, **57**(1), 130–134.

Damanik, D. 2017. Schrödinger operators with dynamically defined potentials. *Ergodic Theory Dynam. Systems*, **37**(6), 1681–1764.

Damanik, D., and Fillman, J. 2022. *One-Dimensional Ergodic Schrödinger Operators – I. General Theory*. Graduate Studies in Mathematics, vol. 221. American Mathematical Society.

Damanik, D., and Fillman, J. 2024. *One-Dimensional Ergodic Schrödinger Operators – II. Specific Classes*. Graduate Studies in Mathematics, vol. 249. American Mathematical Society.

Damle, A., Levitt, A., and Lin, L. 2018. Variational formulation for Wannier functions with entangled band structure. *Preprint*. ArXiv:1801.08572.

Danilov, L. 1990. On the spectrum of the Dirac operator in $\mathbf{R}^n$ with periodic potential. *Teoret. Mat. Fiz.*, **85**(1), 41–53.

Danilov, L. 1995. Resolvent estimates and the spectrum of the Dirac operator with a periodic potential. *Teoret. Mat. Fiz.*, **103**(1), 3–22.

Danilov, L. 1999. On the spectrum of the two-dimensional periodic Dirac operator. *Teoret. Mat. Fiz.*, **118**(1), 3–14.

Danilov, L. 2000a. Absolute continuity of the spectrum of a periodic Dirac operator. *Differ. Uravn.*, **36**(2), 233–240, 287.

Danilov, L. 2000b. On the spectrum of the periodic Dirac operator. *Teoret. Mat. Fiz.*, **124**(1), 3–17.

Danilov, L. 2003a. On the absolute continuity of the spectrum of a periodic Schrödinger operator. *Mat. Zametki*, **73**(1), 49–62.

Danilov, L. 2003b. On the spectrum of the two-dimensional periodic Schrödinger operator. *Teoret. Mat. Fiz.*, **134**(3), 447–459.

Danilov, L. 2005. On the nonexistence of eigenvalues in the spectrum of a generalized two-dimensional periodic Dirac operator. *Algebra i Analiz*, **17**(3), 47–80.

Danilov, L. 2009. On absolute continuity of the spectrum of a periodic magnetic Schrödinger operator. *J. Phys. A*, **42**(27), 275204, 20.

Danilov, L. 2010. On absolute continuity of the spectrum of three- and four-dimensional periodic Schrödinger operators. *J. Phys. A*, **43**(21), 215201, 13.

Danilov, L. 2011. On absolute continuity of the spectrum of a 3D periodic magnetic Dirac operator. *Integr. Equ. Oper. Theory*, **71**(4), 535–556.

Davey, B., Kenig, C., and Wang, J.-N. 2019. On Landis' conjecture in the plane when the potential has an exponentially decaying negative part. *Algebra i Analiz*, **31**(2), 204–226.

Davies, E. B., and Harrell, E. 1987. Conformally flat Riemannian metrics, Schrödinger operators, and semiclassical approximation. *J. Differential Equations*, **66**, 165–188.

Deift, P. A., and Hempel, R. 1986. On the existence of eigenvalues of the Schrödinger operator $H - \lambda W$ in a gap of $\sigma(H)$. *Commun. Math. Phys.*, **103**(3), 461–490.

Derguzov, V. I. 1972a. A mathematical investigation of periodic cylindrical wave guides. I, II. *Vestnik Leningrad. Univ.*, **13 Mat. Meh. Astronom. Vyp. 3**, 32–40, 149; ibid. No. 19 Mat. Meh. Astronom. Vyp. 3 (1972), 14–20, 145.

Derguzov, V. I. 1972b. Solution of an ill-posed problem for a certain class of linear equations with periodic coefficients. Pages 3–28 of: *Problems of Mathematical Analysis, No. 3: Integral and Differential Operators. Differential Equations*. (In Russian.) Izdat. Leningrad. Univ., Leningrad.

Derguzov, V. I. 1975. *Linear Equations with Periodic Coefficients and their Applications to Wave Guide Systems*. (In Russian.) Ph.D. thesis, Leningrad.

Disertori, M., Kirsch, W., Klein, A., Klopp, F., and Rivasseau, V. 2008. *Random Schrödinger Operators*. Panoramas et Synthèses [Panoramas and Syntheses], vol. 25. Société Mathématique de France.

Dixmier, J. 1957. *Les Algèbres d'Opérateurs dans l'Espace Hilbertien (Algèbres de von Neumann)*. Cahiers scientifiques, Fascicule XXV. Gauthier-Villars.

Djakov, P., and Mityagin, B. 2002. Smoothness of Schrödinger operator potential in the case of Gevrey type asymptotics of the gaps. *J. Funct. Anal.*, **195**(1), 89–128.

Djakov, P., and Mityagin, B. 2006. Instability zones of one-dimensional periodic Schrödinger and Dirac operators. *Uspekhi Mat. Nauk*, **61**(4(370)), 77–182.

Djakov, P., and Mityagin, B. 2009. Spectral gaps of Schrödinger operators with periodic singular potentials. *Dyn. Partial Differ. Equ.*, **6**(2), 95–165.

Do, N. T. 2014. On the quantum graph spectra of graphyne nanotubes. *Analysis and Mathematical Physics (2014)*, **5**(1), 39–65.

Do, N. T., and Kuchment, P. 2013. Quantum graph spectra of a graphyne structure. *Nanoscale Systems: Mathematical Modeling, Theory and Applications*, **2**(1), 107–123.

Do, N. T., Kuchment, P., and Sottile, F. 2020. Generic properties of dispersion relations for discrete periodic operators. *J. Math. Phys.*, **61**(10), 103502, 19.

Dubrovin, B. A., Krichever, I. M., and Novikov, S. P. 2001. Integrable systems. I [MR0842910 (87k:58112)]. Pages 177–332 of: *Dynamical Systems, IV*. Encyclopaedia of Mathematical Sciences, vol. 4. Springer.

Dubrovin, B. A., Krichever, I. M., and Novikov, S. P. 2004. *Topological and Algebraic Geometry Methods in Contemporary Mathematical Physics*. Classic Reviews in Mathematics and Mathematical Physics, vol. 2. Cambridge Scientific Publishers.

Dugdale, S. B. 2016. Life on the edge: a beginner's guide to the Fermi surface. *Phys. Scr.*, **91**(5), 053009.

Dunford, N., and Schwartz, J. T. 1988. *Linear Operators. Part II*. Spectral Theory. Selfadjoint Operators in Hilbert Space. With the assistance of William G. Bade and Robert G. Bartle. A reprint of the 1963 original. Wiley Classics Library. John Wiley & Sons Inc.

Dyakin, V. V., and Petrukhnovskiĭ, S. I. 1982. Some geometric properties of Fermi surfaces. *Dokl. Akad. Nauk SSSR*, **264**(5), 1117–1119.

Dyakin, V. V., and Petrukhnovskiĭ, S. I. 1988. On the discreteness of the spectrum of some operator pencils connected with the periodic Schrödinger equation. *Teoret. Mat. Fiz.*, **74**(1), 94–102.

Dynnikov, I. A. 1992. A proof of S. P. Novikov's conjecture for the case of small perturbations of rational magnetic fields. *Uspekhi Mat. Nauk*, **47**(3(285)), 161–162.

Dynnikov, I. A. 1993a. A proof of the conjecture of S. P. Novikov on the semiclassical motion of an electron. *Mat. Zametki*, **53**(5), 57–68.

Dynnikov, I. A. 1993b. S. P. Novikov's problem on the semiclassical motion of an electron. *Uspekhi Mat. Nauk*, **48**(2(290)), 179–180.

Dynnikov, I. A. 1996. Semiclassical electron motion and Novikov conjecture. *Zap. Nauchn. Sem. S.-Peterburg. Otdel. Mat. Inst. Steklov. (POMI)*, **235**(Differ. Geom. Gruppy Li i Mekh. 15-2), 228–234, 306–307.

Dynnikov, I. A. 1997. Semiclassical motion of the electron. A proof of the Novikov conjecture in general position and counterexamples. Pages 45–73 of: *Solitons, Geometry, and Topology: On the Crossroad*. American Mathematical Society Translations: Series 2, vol. 179. American Mathematical Society.

Dynnikov, I. A. 1999. Geometry of the stability zones in the Novikov problem on the semiclassical motion of an electron. *Uspekhi Mat. Nauk*, **54**(1(325)), 21–60.

Eastham, M. S. P. 1973. *The Spectral Theory of Periodic Differential Equations*. Scottish Academic Press Ltd.

Eastham, M. S. P., and Kalf, H. 1982. *Schrödinger-Type Operators with Continuous Spectra*. Research Notes in Mathematics, vol. 65. Pitman.

Egorov, Y. V., and Shubin, M. A. 1998. *Foundations of the Classical Theory of Partial Differential Equations*. Translated from the 1988 Russian original by R. Cooke. Encyclopaedia of Mathematical Sciences [Partial Differential Equations. I], vol. 30. Springer-Verlag.

Ehrenpreis, L. 1970. *Fourier Analysis in Several Complex Variables*. Pure and Applied Mathematics, vol. XVII. Wiley-Interscience Publishers.

Elton, D. 2004. The Bethe–Sommerfield conjecture for the 3-dimensional periodic Landau operator. *Rev. Math. Phys.*, **16**(10), 1259–1290.

Enyanshin, A., and Ivanovskii, A. 2011. Graphene allotropes: stability, structural and electronic properties from DF–TB calculations. *Phys. Status Solidi (B)*, **248**, 1879–1883.

Eremenko, A. 2006. Exceptional values in holomorphic families of entire functions. *Michigan Math. J.*, **54**(3), 687–696.

Eskin, G. 1989. Inverse spectral problem for the Schrödinger equation with periodic vector potential. *Commun. Math. Phys.*, **125**(2), 263–300.

Eskin, G., Ralston, J., and Trubowitz, E. 1984a. Isospectral periodic potentials on $\mathbf{R}^n$. Pages 91–96 of: *Inverse Problems (New York, 1983)*. SIAM-AMS Proceedings, vol. 14. American Mathematical Society.

Eskin, G., Ralston, J., and Trubowitz, E. 1984b. The multidimensional inverse spectral problem with a periodic potential. Pages 45–56 of: *Microlocal Analysis (Boulder, Colorado, 1983)*. Contemporary Mathematics, vol. 27. American Mathematical Society.

Eskin, G., Ralston, J., and Trubowitz, E. 1984c. On isospectral periodic potentials in $\mathbf{R}^n$. II. *Comm. Pure Appl. Math.*, **37**(6), 715–753.

Euler, L. 1743. De integratione aequationum differentialium altiorum graduum. *Miscellanea Berol.*, **7**, 193–242.

Exner, P., and Frank, R. 2007. Absolute continuity of the spectrum for periodically modulated leaky wires in $\mathbb{R}^3$. *Ann. Henri Poincaré*, **8**(2), 241–263.

Exner, P., and Khrabustovskyi, A. 2015. On the spectrum of narrow Neumann waveguide with periodically distributed δ' traps. *J. Phys. A*, **48**(31), 315301, 13.

Exner, P., Kuchment, P., and Winn, B. 2010. On the location of spectral edges in $\mathbb{Z}$-periodic media. *J. Phys. A*, **43**(47), 474022, 8.

Federer, H. 1969. *Geometric Measure Theory*. Grundlehren der Mathematischen Wissenschaften, vol. 153. Springer-Verlag.

Fefferman, C. L., and Weinstein, M. I. 2014. Wave packets in honeycomb structures and two-dimensional Dirac equations. *Commun. Math. Phys.*, **326**(1), 251–286.

Fefferman, C. L., and Weinstein, M. I. 2012. Honeycomb lattice potentials and Dirac points. *J. Amer. Math. Soc.*, **25**(4), 1169–1220.

Fefferman, C. L., Lee-Thorp, J. P., and Weinstein, M. I. 2014. Topologically protected states in one-dimensional continuous systems and Dirac points. *Proc. Natl. Acad. Sci. USA*, **111**(24), 8759–8763.

Fefferman, C. L., Lee-Thorp, J. P., and Weinstein, M. I. 2017. Topologically protected states in one-dimensional systems. *Mem. Amer. Math. Soc.*, **247**(1173), 1–118.

Feldman, J., Knörrer, H., and Trubowitz, E. 1990. The perturbatively stable spectrum of a periodic Schrödinger operator. *Invent. Math.*, **100**(2), 259–300.

Feldman, J., Knörrer, H., and Trubowitz, E. 1991. Perturbatively unstable eigenvalues of a periodic Schrödinger operator. *Comment. Math. Helv.*, **66**(4), 557–579.

Feldman, J., Knörrer, H., and Trubowitz, E. 1992. There is no two-dimensional analogue of the Lame equation. *Math. Ann.*, **294**, 295–324.

Fernandez-Bertolin, A., Roncal, L., and Stan, D. 2024. Landis' conjecture: a survey. *Preprint.* ArXiv:2412.00788

Figotin, A., and Kuchment, P. 1996a. Band-gap structure of spectra of periodic dielectric and acoustic media. I. Scalar model. *SIAM J. Appl. Math.*, **56**(1), 68–88.

Figotin, A., and Kuchment, P. 1996b. Band-gap structure of spectra of periodic dielectric and acoustic media. II. Two-dimensional photonic crystals. *SIAM J. Appl. Math.*, **56**(6), 1561–1620.

Figotin, A., and Kuchment, P. 1998. Spectral properties of classical waves in high-contrast periodic media. *SIAM J. Appl. Math.*, **58**(2), 683–702 (electronic).

Figotin, A., and Vitebskiy, I. 2006. Frozen light in photonic crystals with degenerate band edge. *Phys. Rev. E (3)*, **74**(6), 066613, 17.

Filonov, N. 2001. Second-order elliptic equation of divergence form having a compactly supported solution. *J. Math. Sci. (New York)*, **106**(3), 3078–3086. Function theory and phase transitions.

Filonov, N. 2003. Gaps in the spectrum of the Maxwell operator with periodic coefficients. *Commun. Math. Phys.*, **240**(1-2), 161–170.

Filonov, N., and Kachkovskiy, I. 2018. On the structure of band edges of 2-dimensional periodic elliptic operators. *Acta Math.*, **221**(1), 59–80.

Filonov, N., and Klopp, F. 2004a. Absolute continuity of the spectrum of a Schrödinger operator with a potential which is periodic in some directions and decays in others. *Doc. Math.*, **9**, 107–121.

Filonov, N., and Klopp, F. 2004b. Erratum to: "Absolute continuity of the spectrum of a Schrödinger operator with a potential which is periodic in some directions and decays in others" [Doc. Math. **9** (2004), 107–121]. *Doc. Math.*, **9**, 135–136.

Filonov, N., and Klopp, F. 2005. Absolutely continuous spectrum for the isotropic Maxwell operator and coefficients that are periodic in some directions and decay in others. *Commun. Math. Phys.*, **258**(1), 75–85.

Filonov, N., and Sobolev, A. V. 2004. Absence of the singular continuous component in the spectrum of analytic direct integrals. *Zap. Nauchn. Sem. S.-Peterburg. Otdel. Mat. Inst. Steklov. (POMI)*, **318** (Kraev. Zadachi Mat. Fiz. i Smezh. Vopr. Teor. Funkts. 36 [35]), 298–307, 313.

Firsova, N. E. 1975. The Riemann surface of a quasimomentum, and scattering theory for a perturbed Hill operator. *Zap. Naučn. Sem. Leningrad. Otdel. Mat. Inst. Steklov (LOMI)*, **51**, 183–196, 220.

Firsova, N. E. 2000. On the global quasimomentum in solid state physics. Pages 98–141 of: *Mathematical Methods in Physics (Londrina, 1999)*. World Scientific Publishing.

Fisher, L., Li, W., and Shipman, S. P. 2021. Reducible Fermi surface for multi-layer quantum graphs including stacked graphene. *Commun. Math. Phys.*, **385**(3), 1499–1534.

Floquet, G. 1883. Sur les équations différentielles linéaires à coefficients périodiques. *Ann. Ecole Norm.*, **12**(2), 47–89.

Folland, G. B. 1995. *A Course in Abstract Harmonic Analysis*. Studies in Advanced Mathematics. CRC Press.

Friedlander, L. 1990. On the spectrum of the periodic problem for the Schrödinger operator. *Comm. Partial Differ. Equ.*, **15**(11), 1631–1647.

Friedlander, L. 1991. Erratum to: "On the spectrum of the periodic problem for the Schrödinger operator". *Comm. Partial Differ. Equ.*, **16**(2-3), 527–529.

Friedlander, L. 2002. On the spectrum of a class of second order periodic elliptic differential operators. *Commun. Math. Phys.*, **229**(1), 49–55.

Friedlander, L. 2003. Absolute continuity of the spectra of periodic waveguides. Pages 37–42 of: *Waves in Periodic and Random Media (South Hadley, MA, 2002)*. Contemporary Mathematics, vol. 339. American Mathematical Society.

Froese, R., Herbst, I., Hoffmann-Ostenhof, M., and Hoffmann-Ostenhof, T. 1983. L^2-lower bounds to solutions of one-body Schrödinger equations. *Proc. Roy. Soc. Edinburgh Sect. A*, **95**(1–2), 25–38.

Gasymov, M. G. 1986. Spectral analysis of a class of nonselfadjoint partial differential operators with periodic coefficients. *Dokl. Akad. Nauk SSSR*, **288**(3), 528–530.

Gel′fand, I. M. 1950. Expansion into eigenfunctions of an equation with periodic coefficients. *Dokl. Akad. Nauk SSSR (N.S.)*, **73**, 1117–1120.

Gel′fand, I. M., and Kostyučenko, A. G. 1955. Expansion in eigenfunctions of differential and other operators. *Dokl. Akad. Nauk SSSR (N.S.)*, **103**, 349–352.

Gel′fand, I. M., and Shilov, G. E. 2016. *Generalized functions. Vol. 3*. Theory of Differential Equations. Translated from the 1958 Russian original [MR0106410] by Meinhard E. Mayer, AMS Chelsea Publishing. A reprint of the 1967 English translation [MR0217416].

Gel′fand, I. M., and Vilenkin, N. Ya. 2016. *Generalized Functions. Vol. 4*. Applications of Harmonic Analysis. Translated from the 1961 Russian original [MR0146653] by Amiel Feinstein. AMS Chelsea Publishing. A reprint of the 1964 English translation [MR0173945].

Gérard, C. 1990. Resonance theory for periodic Schrödinger operators. *Bull. Soc. Math. France*, **118**(1), 27–54.

Gérard, C., and Nier, F. 1998a. The Mourre theory for analytically fibered operators. *J. Funct. Anal.*, **152**(1), 202–219.

Gérard, C., and Nier, F. 1998b. Scattering theory for the perturbations of periodic Schrödinger operators. *J. Math. Kyoto Univ.*, **38**(4), 595–634.

Gesztesy, F., and Simon, B. 1988. On a theorem of Deift and Hempel. *Commun. Math. Phys.*, **116**(3), 503–505.

Gesztesy, F., Nichols, R., and Zinchenko, M. 2024. *Sturm–Liouville Operators, their Spectral Theory, and some Applications*. AMS Colloquium Publications, vol. 67. American Mathematical Society.

Gieseker, D., Knörrer, H., and Trubowitz, E. 1991. An overview of the geometry of algebraic Fermi curves. Pages 19–46 of: *Algebraic Geometry: Sundance 1988*. Contemporary Mathematics, vol. 116. American Mathematical Society.

Gieseker, D., Knörrer, H., and Trubowitz, E. 1993. *The Geometry of Algebraic Fermi Curves*. Perspectives in Mathematics, vol. 14. Academic Press Inc.

Glazman, I. M. 1966. *Direct Methods of Qualitative Spectral Analysis of Singular Differential Operators*. Translated from the Russian by the IPST staff. Israel Program for Scientific Translations, Jerusalem, 1965. Daniel Davey & Co., Inc.

Gohberg, I. C., and Kreĭn, M. G. 1957. Fundamental aspects of defect numbers, root numbers and indexes of linear operators. *Uspehi Mat. Nauk (N.S.)*, **12**(2(74)), 43–118. English translation: The basic propositions on defect numbers, root numbers and indices of linear operators, Amer. Math. Soc. Transl. (2) **13** (1960), 185–264.

Gohberg, I. C., and Kreĭn, M. G. 1969. *Introduction to the Theory of Linear Nonselfadjoint Operators*. Translated from the Russian by A. Feinstein. Translations of Mathematical Monographs, vol. 18. American Mathematical Society.

Gordon, C., and Kappeler, T. 1991. On isospectral potentials on tori. *Duke Math. J.*, **63**(1), 217–233.

Gordon, C., and Kappeler, T. 1995. On isospectral potentials on flat tori. II. *Comm. Partial Differ. Equ.*, **20**(3-4), 709–728.

Gordon, C., Guerini, P., Kappeler, T., and Webb, D. 2008. Inverse spectral results on even dimensional tori. *Ann. Inst. Fourier (Grenoble)*, **58**(7), 2445–2501.

Grauert, H. 1958. Analytische Faserungen über holomorph-vollständigen Räumen. *Math. Ann.*, **135**, 163–173.

Grauert, H., and Remmert, R. 2004. *Theory of Stein Spaces*. Translated from the German by Alan Huckleberry, Classics in Mathematics. Springer-Verlag. A reprint of the 1979 translation.

Green, E. 1997. Spectral theory of Laplace–Beltrami operators with periodic metrics. *J. Differential Equations*, **133**(1), 15–29.

Griffiths, P., and Harris, J. 1994. *Principles of Algebraic Geometry*. Wiley Classics Library. John Wiley & Sons, Inc., A reprint of the 1978 original.

Grigoryan, S. A., and Tonev, T. V. 2006. *Shift-Invariant Uniform Algebras on Groups*. Instytut Matematyczny Polskiej Akademii Nauk. Monografie Matematyczne (New Series) [Mathematics Institute of the Polish Academy of Sciences. Mathematical Monographs (New Series)], vol. 68. Birkhäuser Verlag.

Gruber, M. J. 2000. Bloch theory and quantization of magnetic systems. *J. Geom. Phys.*, **34**(2), 137–154.

Gruber, M. J. 2001. Noncommutative Bloch theory. *J. Math. Phys.*, **42**(6), 2438–2465.

Gruber, M. J. 2003. Positive measure spectrum for Schrödinger operators with periodic magnetic fields. *J. Math. Phys.*, **44**(4), 1584–1595.

Grushin, V. V. 2009. Application of the multiparameter theory of perturbations of Fredholm operators to Bloch functions. *Mat. Zametki*, **86**(6), 819–828.

Guillemin, V. 1990. Inverse spectral results on two-dimensional tori. *J. Am. Math. Soc.*, **3**(2), 375–387.

Guillot, J.-C., Ralston, J., and Trubowitz, E. 1988. Semi-classical approximations in solid state physics. Pages 263–269 of: *Partial Differential Equations (Rio de Janeiro, 1986)*. Lecture Notes in Mathematics, vol. 1324. Springer.

Gunning, R. C., and Rossi, H. 2009. *Analytic Functions of Several Complex Variables*. AMS Chelsea Publishing. A reprint of the 1965 original.

Han, D., Kornelson, K., Larson, D., and Weber, E. 2007. *Frames for Undergraduates*. Student Mathematical Library, vol. 40. American Mathematical Society.

Harrison, J. M., Kuchment, P., Sobolev, A., and Winn, B. 2007. On occurrence of spectral edges for periodic operators inside the Brillouin zone. *J. Phys. A*, **40**(27), 7597–7618.

Harrison, W. A., and Webb, M. B. (eds). 1960. *The Fermi Surface*. John Wiley & Sons.

Hasan, M. Z., and Kane, C. L. 2010. Colloquium: Topological insulators. *Rev. Mod. Phys.*, **82**, 3045–3067.

Heebner, J. E., Boyd, R. W., and Park, Q. 2002. SCISSOR Solitons & other propagation effects in microresonator modified waveguides. *JOSA B*, **19**, 722–731.

Helffer, B., and Hoffmann-Ostenhof, T. 2003. Spectral theory for periodic Schrödinger operators with reflection symmetries. *Commun. Math. Phys.*, **242**(3), 501–529.

Helffer, B., and Mohamed, A. 1998. Asymptotic of the density of states for the Schrödinger operator with periodic electric potential. *Duke Math. J.*, **92**(1), 1–60.

Helffer, B., and Sjöstrand, J. 1989. Opérateurs de Schrödinger avec champs magnétiques faibles et constants. Pages Exp. No. XII, 11 of: *Séminaire sur les Équations aux Dérivées Partielles, 1988–1989*. École Polytech., Palaiseau.

Helffer, B., and Sjöstrand, J. 1990. On diamagnetism and de Haas–van Alphen effect. *Ann. Inst. H. Poincaré Phys. Théor.*, **52**(4), 303–375.

Hempel, R. 1989. On the asymptotic distribution of the eigenvalue branches of the Schrödinger operator $H \pm \lambda W$ in a spectral gap of *H*. *J. Reine Angew. Math.*, **399**, 38–59.

Hempel, R. 1992. Eigenvalue asymptotics related to impurities in crystals. In *Méthodes Semi-Classiques, Vol. 2 (Nantes, 1991)*. *Astérisque*, **210**, 8, 183–196.

Hempel, R. 2008. Eigenvalues of Schrödinger operators in gaps of the essential spectrum – an overview. Pages 393–407 of: *Integrable Systems and Random Matrices*. Contemporary Mathematics, vol. 458. American Mathematical Society.

Hempel, R., and Herbst, I. 1995. Bands and gaps for periodic magnetic Hamiltonians. Pages 175–184 of: *Partial Differential Operators and Mathematical Physics (Holzhau, 1994)*. Operator Theoretical Methods and Applications, vol. 78. Birkhäuser.

Hempel, R., and Laitenberger, J. 1994. Schrödinger operators with strong local magnetic perturbations: existence of eigenvalues in gaps of the essential spectrum. Pages 13–18 of: *Mathematical Results in Quantum Mechanics (Blossin, 1993)*. Operator Theoretical Methods and Applications, vol. 70. Birkhäuser.

Hempel, R., and Levendorskiĭ, S. Z. 1998. On eigenvalues in gaps for perturbed magnetic Schrödinger operators. *J. Math. Phys.*, **39**(1), 63–78.

Hempel, R., and Post, O. 2003. Spectral gaps for periodic elliptic operators with high contrast: an overview. Pages 577–587 of: *Progress in Analysis, Vol. I, II (Berlin, 2001)*. World Scientific Publishing.

Hess, P. 1987. Positive solutions of periodic-parabolic problems and stability. Pages 129–136 of: *Nonlinear Parabolic Equations: Qualitative Properties of Solutions (Rome, 1985)*. Pitman Research Notes in Mathematics Series, vol. 149. Longman Scientific & Technical Publisher.

Hislop, P. D., and Sigal, I. M. 1996. *Introduction to Spectral Theory*. With applications to Schrödinger operators. Applied Mathematical Sciences, vol. 113. Springer-Verlag.

Hitrik, M., and Polterovich, I. 2003. Regularized traces and Taylor expansions for the heat semigroup. *J. London Math. Soc.*, **68**, 402–418.

Hoefer, M. A., and Weinstein, M. I. 2011. Defect modes and homogenization of periodic Schrödinger operators. *SIAM J. Math. Anal.*, **43**(2), 971–996.

Hörmander, L. 1990. *An Introduction to Complex Analysis in Several Variables*. 3rd ed. North-Holland Mathematical Library, vol. 7. North-Holland Publishing Co.

Hörmander, L. 2005. *The Analysis of Linear Partial Differential Operators. II*. Differential operators with constant coefficients, Classics in Mathematics. Springer-Verlag. A reprint of the 1983 original.

Hu, Y. et al. 1998. Rich nature of Van Hove singularities in Kagome superconductor CsV3Sb5. *Nat. Commun.*, **13**, 2022.

Husemoller, D. 1975. *Fibre Bundles*. 2nd ed. Graduate Texts in Mathematics, No. 20. Springer-Verlag.

Iakubovich, V. A., and Starzhinskii, V. M. 1975. *Linear Differential Equations with Periodic Coefficients. 1, 2*. Translated from Russian by D. Louvish. Halsted Press. [John Wiley & Sons; Israel Program for Scientific Translations.]

Igoshev, P. A. and Irkhin, V. Y. 2022. Giant density-of-states van Hove singularities in the face-centered cubic lattice. *Physics Letters A*, **438**, 128107.

Ivrii, V. 1998. *Microlocal Analysis and Precise Spectral Asymptotics*. Springer Monographs in Mathematics. Springer-Verlag.

Ivrii, V. 1999. Accurate spectral asymptotics for periodic operators. Pages Exp. No. V, 11 of: *Journées "Équations aux Dérivées Partielles" (Saint-Jean-de-Monts, 1999)*. Univ. Nantes, Nantes.

Ivrii, V. 2019a. Bethe-Sommerfeld Conjecture in Semiclassical Settings. Pages 619–639 of: *Microlocal Analysis, Sharp Spectral Asymptotics and Applications V*. Springer International Publishing.

Ivrii, V. 2019b. *Microlocal Analysis, Sharp Spectral Asymptotics and Applications. I*. Semiclassical microlocal analysis and local and microlocal semiclassical asymptotics. Springer.

Ivrii, V. 2019c. *Microlocal Analysis, Sharp Spectral Asymptotics and Applications. II*. Functional methods and eigenvalue asymptotics. Springer.

Ivrii, V. 2019d. *Microlocal Analysis, Sharp Spectral Asymptotics and Applications. III*. Magnetic Schrödinger operator 1. Springer.

Ivrii, V. 2019e. *Microlocal Analysis, Sharp Spectral Asymptotics and Applications. IV*. Magnetic Schrödinger operator 3. Springer.

Ivrii, V. 2019f. *Microlocal Analysis, Sharp Spectral Asymptotics and Applications. V. Applications to Quantum Theory and Miscellaneous Problems*. Springer.

Ivrii, V. 2021. Complete semiclassical spectral asymptotics for periodic and almost periodic perturbations of constant operators. Pages 183–201 of: *Differential Equations on Manifolds and Mathematical Physics – Dedicated to the Memory of Boris Sternin*. Trends in Mathematics. Birkhäuser/Springer.

Jikov, V. V., Kozlov, S. M., and Oleĭnik, O. A. 1994. *Homogenization of Differential Operators and Integral Functionals*. Translated from the Russian by G. A. Yosifian [G. A. Iosif′yan]. Springer-Verlag.

Joannopoulos, J. D., Johnson, S., Meade, R. D., and Winn, J. N. 2008. *Photonic Crystals: Molding the Flow of Light*. 2nd ed. Princeton University Press.

John, S. 1987. Strong localization of photons in certain disordered dielectric superlattices. *Phys. Rev. Lett.*, **58**(23), 2486–2489.

Johnson, R., and Moser, J. 1982. The rotation number for almost periodic potentials. *Commun. Math. Phys.*, **84**(3), 403–438.

Johnson, R., and Moser, J. 1983. Erratum: "The rotation number for almost periodic potentials" [Comm. Math. Phys. **84** (1982), no. 3, 403–438; MR0667409 (83h:34018)]. *Commun. Math. Phys.*, **90**(2), 317–318.

Julia, G. 1926. Sur le domaine d'existence d'une fonction implicite définie par une relation entière $G(x, y) = 0$. *Bull. Soc. Math. France*, **54**, 26–37.

Kachkovskiĭ, I., and Filonov, N. 2009. Absolute continuity of the spectrum of a periodic Schrödinger operator in a multidimensional cylinder. *Algebra i Analiz*, **21**(1), 133–152.

Kachkovskiĭ, I., and Filonov, N. 2010. Absolute continuity of the spectrum of the periodic Schrödinger operator in a layer and in a smooth cylinder. *Zap. Nauchn. Sem. S.-Peterburg. Otdel. Mat. Inst. Steklov. (POMI)*, **385** (Kraevye Zadachi Matematicheskoĭ Fiziki i Smezhnye Voprosy Teorii Funktsiĭ. 41), 69–81, 235.

Kamotski, I. V., and Smyshlyaev, V. P. 2018. Localized modes due to defects in high contrast periodic media via two-scale homogenization. *J. Math. Sci. (N.Y.)*, **232**(3, Problems in mathematical analysis. No. 92 (Russian)), 349–377.

Kappeler, T., Schaad, B., and Topalov, P. 2012. Asymptotics of spectral quantities of Schrödinger operators. Pages 243–284 of: *Spectral Geometry*. Proceedings of Symposia in Pure Mathematics, vol. 84. American Mathematical Society.

Karpeshina, Y. 1995. Perturbation formulae for the Schrödinger operator with a periodic potential near planes of diffraction. *Math. Res. Lett.*, **2**(1), 59–74.

Karpeshina, Y. 1997. *Perturbation Theory for the Schrödinger Operator with a Periodic Potential*. Lecture Notes in Mathematics, vol. 1663. Springer-Verlag.

Karpeshina, Y. 2000. On the density of states for the periodic Schrödinger operator. *Ark. Mat.*, **38**(1), 111–137.

Karpeshina, Y., and Lee, Y.-R. 2008. Absolutely continuous spectrum of a polyharmonic operator with a limit periodic potential in dimension two. *Comm. Partial Differ. Equ.*, **33**(7-9), 1711–1728.

Karpeshina, Y. E. 1989. Perturbation theory for Schroedinger operator with a periodic potential. Pages 130–145 of: *Schrödinger Operators, Standard and Nonstandard (Dubna, 1988)*. World Scientific Publishing.

Karpeshina, Y. E. 1990. Perturbation theory for the Schrödinger operator with periodic potential. *Trudy Mat. Inst. Steklov.*, **188**, 88–116, 191. Translated in Proceedings of the Steklov Institute of Mathematics 1991, no. 3, 109–145, Boundary value problems of mathematical physics, 14 (Russian).

Kato, T. 1995. *Perturbation Theory for Linear Operators*. Classics in Mathematics. Springer-Verlag. Reprint of the 1980 edition.

Katsnelson, M. I. 2012. *Graphene. Carbon in Two Dimensions*. Cambridge University Press.

Kenig, C., Silvestre, L., and Wang, Jenn-Nan. 2015. On Landis' conjecture in the plane. *Comm. Partial Differ. Equ.*, **40**(4), 766–789.

Kha, M. 2018. Green's function asymptotics of periodic elliptic operators on abelian coverings of compact manifolds. *J. Funct. Anal.*, **274**(2), 341–387.

Kha, M., and Kuchment, P. 2021. *Liouville–Riemann–Roch theorems on abelian coverings*. Lecture Notes in Mathematics, vol. 2245. Springer.

Kha, M., Kuchment, P., and Raich, A. 2017. Green's function asymptotics near the internal edges of spectra of periodic elliptic operators. Spectral gap interior. *J. Spectr. Theory*, **7**(4), 1171–1233.

Kha, M., and Lin, V. 2019. Polynomial-like elements in vector spaces with group actions. Pages 193–217 of: *Differential Equations, Mathematical Physics, and Applications: Selim Grigorievich Krein Centennial*. Contemporary Mathematics, vol. 734. American Mathematical Society.

Khrabustovskyi, A. 2012. Periodic Riemannian manifold with preassigned gaps in spectrum of Laplace-Beltrami operator. *J. Differential Equations*, **252**(3), 2339–2369.

Khrabustovskyi, A. 2013. Periodic elliptic operators with asymptotically preassigned spectrum. *Asymptot. Anal.*, **82**(1-2), 1–37.

Khrabustovskyi, A. 2014. Opening up and control of spectral gaps of the Laplacian in periodic domains. *J. Math. Phys.*, **55**(12), 121502, 23.

Khrabustovskyi, A. 2015. Asymptotically predefined spectral gaps for the Neumann Laplacian in periodic domains. Pages 267–274 of: *Mathematical Results in Quantum Mechanics*. World Scientific Publishing.

Khrabustovskyi, A., and Khruslov, E. 2015. Gaps in the spectrum of the Neumann Laplacian generated by a system of periodically distributed traps. *Math. Methods Appl. Sci.*, **38**(1), 11–26.

Kirsch, W., and Simon, B. 1987. Comparison theorems for the gap of Schrödinger operators. *J. Funct. Anal.*, **75**(2), 396–410.

Kiselev, A., and Last, Y. 2000. Solutions, spectrum, and dynamics for Schrödinger operators on infinite domains. *Duke Math. J.*, **102**(1), 125–150.

Kitaev, A. 2009. Periodic table for topological insulators and superconductors. Pages 22–30 of: *AIP Conference Proceedings*, vol. 1134.

Kittel, C. 1976. *Introduction to Solid State Physics*. Wiley.

Klopp, F., and Ralston, J. 2000. Endpoints of the spectrum of periodic operators are generically simple. *Methods Appl. Anal.*, **7**(3), 459–463.

Knörrer, H., and Trubowitz, E. 1990. A directional compactification of the complex Bloch variety. *Comment. Math. Helv.*, **65**(1), 114–149.

Knox, R. S., and Gold, A. 1964. *Symmetry in the Solid State*. W. A. Benjamin, Inc.

Kobayashi, T., Ono, K., and Sunada, T. 1989. Periodic Schrödinger operators on a manifold. *Forum Math.*, **1**(1), 69–79.

Kohn, W. 1959. Analytic properties of Bloch waves and Wannier functions. *Phys. Rev. (2)*, **115**, 809–821.

Kohn, W. 1973. Construction of Wannier functions and applications to energy bands. *Phys. Rev. B (3)*, **7**(10), 4388–4398.

Kondrat′ev, V. A., and Landis, E. M. 1988. Qualitative properties of solutions of some nonlinear elliptic equations. Pages 100–103, 267 of: *Functional and Numerical Methods in Mathematical Physics (Russian)*. "Naukova Dumka".

Korotyaev, E., and Lobanov, I. 2007. Schrödinger operators on zigzag nanotubes. *Ann. Henri Poincaré*, **8**(6), 1151–1176.

Krantz, S., and Parks, H. 2002. *A Primer of Real Analytic Functions*. 2nd ed. Birkhäuser Advanced Texts: Basler Lehrbücher. [Birkhäuser Advanced Texts: Basel Textbooks]. Birkhäuser Boston, Inc., Boston, MA.

Krantz, S. G. 1999. *A Panorama of Harmonic Analysis*. Carus Mathematical Monographs, vol. 27. Mathematical Association of America, Washington, DC.

Krein, S. G. 1982. *Linear Equations in Banach Spaces*. Boston, Mass.: Birkhäuser. Translated from the Russian by A. Iacob. With an introduction by I. Gohberg.

Krein, S. G. et al. 1972. *Funktsionalnyi analiz*. Izdat. “Nauka”, Moscow. Edited by S. G. Kreĭn, 2nd ed., revised and augmented, Mathematical Reference Library.

Krupchyk, K., and Uhlmann, G. 2017. Absolute continuity of the periodic Schrödinger operator in transversal geometry. *J. Eur. Math. Soc. (JEMS)*, **19**(2), 531–550.

Krymskii, S. T., and Filonov, N. D. 2023. On the rate of decay at infinity for solutions to the Schrödinger equation in a half-cylinder. *St. Petersburg Math. J.*, **34**(1), 79–92. Translated from Algebra i Analiz **34** (2022), No. 1.

Krymskii, S. T., Logunov, A., and Pagano, F. 2025. Eigenfunctions with double exponential rate of localization. Preprint. ArXiv:2501.15354.

Kuchment, P. 1982. Floquet theory for partial differential equations. *Uspekhi Mat. Nauk.* **37**(4), 1–60; Translation in *Russian Math. Surveys* **37**(4(226)), 3–52, 240.

Kuchment, P. 1986a. On periodic perturbations which preserve the dispersion relation of a differential operator. *Differencial’nye Uravneniya*, **22**(9), 1641–1642.

Kuchment, P. 1986b. Periodic perturbations that preserve the law of dispersion of a differential operator. *Differentsial′nye Uravneniya*, **22**(9), 1641–1642, 1655.

Kuchment, P. 1991. On the Floquet theory of periodic difference equations. Pages 201–209 of: *Geometrical and Algebraical Aspects in Several Complex Variables (Cetraro, 1989)*. Sem. Conf., vol. 8. Rende: EditEl.

Kuchment, P. 1993. *Floquet Theory for Partial Differential Equations*. Operator Theory: Advances and Applications, vol. 60. Birkhäuser Verlag.

Kuchment, P. 2001. The mathematics of photonic crystals. Pages 207–272 of: *Mathematical Modeling in Optical Science*, Bao, G., Cowsar, L., and Masters, W. (eds). Frontiers in Applied Mathematics, vol. 22. SIAM.

Kuchment, P. 2005. Quantum graphs. II. Some spectral properties of quantum and combinatorial graphs. *J. Phys. A*, **38**(22), 4887–4900.

Kuchment, P. 2009. Tight frames of exponentially decaying Wannier functions. *J. Phys. A*, **42**(2), 025203, 16.

Kuchment, P. 2015. Introduction to Periodic Operators. Video at `www.newton.ac.uk/event/pepw01/timetable`.

Kuchment, P. 2016. An overview of periodic elliptic operators. *Bull. Amer. Math. Soc. (N.S.)*, **53**(3), 343–414.

Kuchment, P. 2027. *Periodic Differential Operators. Generalizations and Applications*. In preparation.

Kuchment, P., and Kunyansky, L. 1999. Spectral properties of high contrast band-gap materials and operators on graphs. *Experiment. Math.*, **8**(1), 1–28.

Kuchment, P., and Levendorskiî, S. 2002. On the structure of spectra of periodic elliptic operators. *Trans. Amer. Math. Soc.*, **354**(2), 537–569.

Kuchment, P., and Pinchover, Y. 2001. Integral representations and Liouville theorems for solutions of periodic elliptic equations. *J. Funct. Anal.*, **181**(2), 402–446.

Kuchment, P., and Pinchover, Y. 2007. Liouville theorems and spectral edge behavior on abelian coverings of compact manifolds. *Trans. Amer. Math. Soc.*, **359**(12), 5777–5815.

Kuchment, P., and Post, O. 2007. On the spectra of carbon nano-structures. *Commun. Math. Phys.*, **275**(3), 805–826.

Kuchment, P., and Vainberg, B. 1998. On embedded eigenvalues of perturbed periodic Schrödinger operators. Pages 67–75 of: *Spectral and Scattering Theory (Newark, DE, 1997)*. Plenum.

Kuchment, P., and Vainberg, B. 2000. On absence of embedded eigenvalues for Schrödinger operators with perturbed periodic potentials. *Comm. Partial Differ. Equ.*, **25**(9-10), 1809–1826.

Kuchment, P., and Vainberg, B. 2006. On the structure of eigenfunctions corresponding to embedded eigenvalues of locally perturbed periodic graph operators. *Commun. Math. Phys.*, **268**(3), 673–686.

Kuiper, N. 1965. The homotopy type of the unitary group of Hilbert space. *Topology*, **3**, 19–30.

Landis, E. M. 1963a. On the growth and decrease of solutions of linear elliptic equations of second order. Pages 151–154 of: *Outlines Joint Sympos. Partial Differential Equations (Novosibirsk, 1963)*. Academy of Sciences. USSR Siberian Branch.

Landis, E. M. 1963b. Some questions in the qualitative theory of second-order elliptic equations (case of several independent variables). *Uspehi Mat. Nauk*, **18**(1 (109)), 3–62.

Lapin, I. S. 2003. One version of the Bethe-Sommerfeld conjecture. *J. Math. Sci. (N. Y.)*, **117**(3), 4157–4166. Nonlinear problems and function theory.

Lawson, Jr., H. B., and Michelsohn, M.-L. 1989. *Spin geometry*. Princeton Mathematical Series, vol. 38. Princeton University Press.

Lee, M. 2016. Dirac Cones for Point Scatterers on a Honeycomb Lattice. *SIAM J. Math. Anal.*, **48**(2), 95–102.

Levendorskiĭ, S. Z. 1985. Spectral asymptotics of differential operators with operator coefficients and some applications. *Dokl. Akad. Nauk SSSR*, **280**(6), 1303–1306.

Levendorskiĭ, S. Z. 1995a. Precise spectral asymptotics for the Schrödinger operator with perturbed periodic potential. *Preprint*.

Levendorskiĭ, S. Z. 1995b. Lower bounds for the number of eigenvalue branches for the Schrödinger operator $H - \lambda W$ in a gap of H: the case of indefinite W. *Comm. Partial Differ. Equ.*, **20**(5-6), 827–854.

Levendorskiĭ, S. Z. 1996. Magnetic Floquet theory and spectral asymptotics for Schrödinger operators. *Funktsional. Anal. i Prilozhen.*, **30**(4), 77–80.

Levendorskiĭ, S. Z. 1997. Floquet's theory for the Schrödinger operator with perturbed periodic potential, and exact spectral asymptotics. *Dokl. Akad. Nauk*, **355**(1), 21–24.

Levin, B. Y. 1996. *Lectures on Entire Functions*. In collaboration with and with a preface by Yu. Lyubarskii, M. Sodin and V. Tkachenko, Translated from the Russian manuscript by Tkachenko. Translations of Mathematical Monographs, vol. 150. American Mathematical Society.

Levitin, M., Sobolev, A. V., and Sobolev, D. 2010. On the near periodicity of eigenvalues of Toeplitz matrices. Pages 115–125 of: *Operator theory and its applications*. American Mathematical Society Translations: Series 2, vol. 231. American Mathematical Society.

Li, W., and Shipman, S. P. 2020. Irreducibility of the Fermi surface for planar periodic graph operators. *Lett. Math. Phys.*, **110**(9), 2543–2572.

Lin, V., and Pinchover, Y. 1994. Manifolds with group actions and elliptic operators. *Mem. Amer. Math. Soc.*, **112**(540), vi+78.

Lions, J.-L., and Magenes, E. 1972. *Non-Homogeneous Boundary Value Problems and Applications. Vol. I*. Translated from the French by P. Kenneth. Grundlehren der mathematischen Wissenschaften, vol. 181. Springer-Verlag.

Liu, J., N'Guérékata, G., and Minh, N. V. 2008. *Topics on Stability and Periodicity in Abstract Differential Equations*. Series on Concrete and Applicable Mathematics, vol. 6. World Scientific Publishing Co. Pte. Ltd.

Liu, W. 2022. Irreducibility of the Fermi variety for discrete periodic Schrödinger operators and embedded eigenvalues. *Geom. Funct. Anal.*, **32**(1), 1–30.

Liu, W. 2023a. Fermi isospectrality of discrete periodic Schrödinger operators with separable potentials on $\mathbb{Z}^2$. *Commun. Math. Phys.*, **399**(2), 1139–1149.

Liu, W. 2023b. Floquet isospectrality for periodic graph operators. *J. Differential Equations*, **374**, 642–653.

Liu, W. 2023c. Proof of geometric Borg's theorem in arbitrary dimensions. Preprint. ArXiv:2306.16412v1.

Liu, W. 2024. Bloch varieties and quantum ergodicity for periodic graph operators. *J. Anal. Math.*, **153**(2), 671–681.

Lledó, F., and Post, O. 2007. Generating spectral gaps by geometry. Pages 159–169 of: *Prospects in Mathematical Physics*. Contemporary Mathematics, vol. 437. American Mathematical Society.

Lledó, F., and Post, O. 2008. Existence of spectral gaps, covering manifolds and residually finite groups. *Rev. Math. Phys.*, **20**(2), 199–231.

Logunov, A., and Malinnikova, E. 2018. Quantitative propagation of smallness for solutions of elliptic equations. Pages 2391–2411 of: *Proceedings of the International Congress of Mathematicians – Rio de Janeiro 2018. Vol. III. Invited lectures*. World Scientific Publishing.

Logunov, A., and Malinnikova, E. 2020. Lecture notes on quantitative unique continuation for solutions of second order elliptic equations. Pages 1–33 of: *Harmonic Analysis and Applications*. IAS/Park City Mathematics Series, vol. 27. American Mathematical Society.

Lojasiewicz, S. 1974. Stratification des ensembles analytiques avec les propriétés (A) et (B) de Whitney. Pages 116–130. "Agora Mathematica", No. 1 of: *Fonctions Analytiques de Plusieurs Variables et Analyse Complexe (Colloq. Internat. C.N.R.S. No. 208, Paris, 1972)*. Gauthier-Villars.

Lojasiewicz, S. 1991. *Introduction to Complex Analytic Geometry*. Translated from the Polish by Maciej Klimek. Birkhäuser Verlag.

Ludewig, M., and Thiang, G. C. 2020. Good Wannier bases in Hilbert modules associated to topological insulators. *J. Math. Phys.*, **61**(6), 061902, 22.

Lyapunov, A. M. 1896. Sur une série relat ive a la theorie des equations differentielles lineaires a coefficients periodiques. *Compt. Rend.*, **123**(26), 1248–1252.

Lyapunov, A. M. 1899. Sur une equation transcendante et les equations differentielles lineaires du second ordre a coefficients periodiques. *Compt. Rend.*, **128**(18), 1085–1088.

Mackey, G. W. 1978. Harmonic analysis as the exploitation of symmetry – a historical survey. History of Analysis (Proc. Conf., Rice Univ., Houston, Tex., 1977). *Rice Univ. Stud.*, **64**(2–3), 73–228.

Magnus, W., and Winkler, S. 1979. *Hill's Equation*. Corrected reprint of the 1966 edition. Dover Publications, Inc.

Malgrange, B. 1955–1956. Existence et approximation des solutions des équations aux dérivées partielles et des équations de convolution. *Ann. Inst. Fourier, Grenoble*, **6**, 271–355.

Malgrange, B. 1963. Sur les systèmes différentiels à coefficients constants. Pages 113–122 of: *Les Équations aux Dérivées Partielles (Paris, 1962)*. Editions du Centre National de la Recherche Scientifique.

Malgrange, B. 1984. *Lectures on the Theory of Functions of Several Complex Variables*. Reprint of the 1958 edition, Notes by Raghavan Narasimhan. Tata Institute of Fundamental Research Lectures on Mathematics and Physics, vol. 13.

Marcelli, G., Moscolari, M., and Panati, G. 2023. Localization of generalized Wannier bases implies Chern triviality in non-periodic insulators. *Ann. Henri Poincaré*, **24**(3), 895–930.

Marchenko, V. A. 2011. *Sturm–Liouville Operators and Applications*. Revised edn. AMS Chelsea Publishing.

Marchenko, V. A., and Ostrovsky, I. V. 1975. A characterization of the spectrum of the Hill operator. *Mat. Sb. (N.S.) English transl. in Math. USSR-Sb. 26 (175)*, **97(139)**(4(8)), 540–606, 633–634.

Marchenko, V. A., and Ostrovsky, I. V. 1980. Approximation of periodic potentials by finite zone potentials. *Vestnik Khar′kov. Gos. Univ.*, **205**, 4–40, 139.

Marchenko, V. A., and Ostrovsky, I. V. 1988. Corrections to the article: "Approximation of periodic by finite-zone potentials" [Selecta Math. Soviet. **6**(2), 101–136; see MR0910538 (88f:00011)]. *Selecta Math. Soviet.*, **7**(1), 99–100.

Marx, C. A., and Jitomirskaya, S. 2017. Dynamics and spectral theory of quasi-periodic Schrödinger-type operators. *Ergodic Theory and Dynamical Systems*, **37**(8), 2353–2393.

Marzari, N., and Vanderbilt, D. 1997. Maximally localized generalized Wannier functions for composite energy bands. *Phys. Rev. B*, **56**(20), 12847–12865.

Marzari, N., Souza, I., I, and Vanderbilt, D. 2003. An introduction to maximally-localized Wannier functions. *Highlight of the Month, Psi-K Newsletter*, **57**(4), 129–168.

Marzari, N., Mostofi, A., Yates, J., Souza, I., and Vanderbilt, D. 2012. Maximally localized Wannier functions: theory and applications. *Rev. Mod. Phys.*, **84**, 1419–1475.

McKean, H. P., and Trubowitz, E. 1976. Hill's operator and hyperelliptic function theory in the presence of infinitely many branch points. *Comm. Pure Appl. Math.*, **29**(2), 143–226.

McKean, H. P., and van Moerbeke, P. 1975. The spectrum of Hill's equation. *Invent. Math.*, **30**(3), 217–274.

Melrose, R. B. 1993. *The Atiyah–Patodi–Singer Index Theorem*. Research Notes in Mathematics, vol. 4. A. K. Peters, Ltd.

Meshkov, V. 1991. On the possible rate of decrease at infinity of the solutions of second-order partial differential equations. *Mat. Sb.*, **182**(3), 364–383. (In Russian.) English transl. in Math. USSR-Sbornik.

Milton, G. W. 2002. *The Theory of Composites*. Cambridge Monographs on Applied and Computational Mathematics, vol. 6. Cambridge University Press.

Milton, G. W. (ed). 2016. *Extending the Theory of Composites to Other Areas of Science*. `www.amazon.com/Extending-Theory-Composites-Other-Science/dp/1483569195`.

Mitjagin, B. S. 1970. The homotopy structure of a linear group of a Banach space. *Uspehi Mat. Nauk*, **25**(5(155)), 63–106.

Mityagin, B. S. 2020. The Zero Set of a Real Analytic Function. *Mat. Zametki*, **107**(3), 473–475.

Mizohata, S. 1973. *The Theory of Partial Differential Equations*. Translated from the Japanese by Katsumi Miyahara. Cambridge University Press.

Mohamed, A. 1997. Asymptotic of the density of states for the Schrödinger operator with periodic electromagnetic potential. *J. Math. Phys.*, **38**(8), 4023–4051.

Molchanov, S., and Vainberg, B. 2004. Slowdown of the wave packages in finite slabs of periodic media. *Waves Random Media*, **14**(3), 411–423.

Molchanov, S., and Vainberg, B. 2005. Slowing down of the wave packets in quantum graphs. *Waves Random Complex Media*, **15**(1), 101–112.

Molchanov, S., and Vainberg, B. 2006. Slowing down and reflection of waves in truncated periodic media. *J. Funct. Anal.*, **231**(2), 287–311.

Monaco, D., and Panati, G. 2014. Topological invariants of eigenvalue intersections and decrease of Wannier functions in graphene. *J. Stat. Phys.*, **155**(6), 1027–1071.

Morame, A. 2000. The absolute continuity of the spectrum of Maxwell operator in a periodic media. *J. Math. Phys.*, **41**(10), 7099–7108.

Morozov, S., Parnovski, L., and Pchelintseva, I. 2010. Lower bound on the density of states for periodic Schrödinger operators. Pages 161–171 of: *Operator Theory and its Applications*. American Mathematical Society Translations: Series 2, vol. 231. American Mathematical Society.

Morozov, S., Parnovski, L., and Shterenberg, R. 2014. Complete asymptotic expansion of the integrated density of states of multidimensional almost-periodic pseudo-differential operators. *Ann. Henri Poincaré*, **15**(2), 263–312.

Moscolari, M., and Panati, G. 2023. Ultra-generalized Wannier bases: are they relevant to topological transport? *J. Math. Phys.*, **64**(7), Paper No. 071901, 12.

Mostofi, A. A., Yates, J. R., Lee, Y.-S., Souza, I., Vanderbilt, D., and Marzari, N. 2008. Wannier90: a tool for obtaining maximally-localised Wannier functions. *Comput. Phys. Commun.*, **178**(9), 685–699.

Mostofi, A. A., Yates, J. R., Giovanni, P., Lee, Y.-S., Souza, I., Vanderbilt, D., and Marzari, N. 2014. An updated version of Wannier90: A tool for obtaining maximally-localised Wannier functions. *Comput. Phys. Commun.*, **185**(8), 2309–2310.

Murata, M., and Tsuchida, T. 2006. Asymptotics of Green functions and the limiting absorption principle for elliptic operators with periodic coefficients. *J. Math. Kyoto Univ.*, **46**(4), 713–754.

Narasimhan, R. 1966. *Introduction to the theory of analytic spaces*. Lecture Notes in Mathematics, No. 25. Springer-Verlag.

Nenciu, G. 1983. Existence of the exponentially localised Wannier functions. *Commun. Math. Phys.*, **91**(1), 81–85.

Nielsen, O. 1980. *Direct Integral Theory*. Lecture Notes in Pure and Applied Mathematics, vol. 61. Marcel Dekker, Inc.

Novikov, S. P. 1981. Bloch functions in the magnetic field and vector bundles. Typical dispersion relations and their quantum numbers. *Dokl. Akad. Nauk SSSR*, **257**(3), 538–543.

Novikov, S. P. 1983. Two-dimensional Schrödinger operators in periodic fields. Pages 3–32 of: *Current Problems in Mathematics, Vol. 23*. Itogi Nauki i Tekhniki. Akad. Nauk SSSR, Vsesoyuz. Inst. Nauchn. i Tekhn. Inform.

Novikov, S. P. 1994. *Solitons and Geometry*. Lezioni Fermiane. [Fermi Lectures]. Published for the Scuola Normale Superiore, Pisa.

Novikov, S. P. 1999. Levels of quasiperiodic functions on a plane, and Hamiltonian systems. *Uspekhi Mat. Nauk*, **54**(5(329)), 147–148.

Novikov, S. P., Manakov, S., Pitaevskiĭ, L., and Zakharov, V. 1984. *Theory of Solitons*. The inverse scattering method. Translated from the Russian. Contemporary Soviet Mathematics. Consultants Bureau [Plenum].

Ong, B.-S. 2006. *Spectral Problems of Optical Waveguides and Quantum Graphs*. Ph.D. thesis, Texas A&M University.

Palais, R. S. 1965. *Seminar on the Atiyah-Singer index theorem*. With contributions by M. F. Atiyah, A. Borel, E. E. Floyd, R. T. Seeley, W. Shih, and R. Solovay. Annals of Mathematics Studies, No. 57. Princeton University Press.

Palamodov, V. P. 1970. *Linear Differential Operators with Constant Coefficients*. Translated from the Russian by A. A. Brown. Grundlehren der mathematischen Wissenschaften, vol. 168. Springer-Verlag.

Panati, G. 2007. Triviality of Bloch and Bloch–Dirac bundles. *Ann. Henri Poincaré*, **8**(5), 995–1011.

Panati, G., and Pisante, A. 2013. Bloch bundles, Marzari–Vanderbilt functional and maximally localized Wannier functions. *Commun. Math. Phys.*, **322**(3), 835–875.

Papanicolaou, V. 2021. Examples of fourth-order scattering-type operators with embedded eigenvalues in their continuous spectra. Preprint, ArXiv::2110.00739.

Parnovski, L. 2008. Bethe–Sommerfeld conjecture. *Ann. Henri Poincaré*, **9**(3), 457–508.

Parnovski, L., and Shterenberg, R. 2012. Complete asymptotic expansion of the integrated density of states of multidimensional almost-periodic Schrödinger operators. *Ann. Math. (2)*, **176**(2), 1039–1096.

Parnovski, L., and Shterenberg, R. 2017. Perturbation theory for spectral gap edges of 2D periodic Schrödinger operators. *J. Funct. Anal.*, **273**(1), 444–470.

Parnovski, L., and Sidorova, N. 2010. Critical dimensions for counting lattice points in Euclidean annuli. *Math. Model. Nat. Phenom.*, **5**(4), 293–316.

Parnovski, L., and Sobolev, A. 2001a. Lattice points, perturbation theory and the periodic polyharmonic operator. *Ann. Henri Poincaré*, **2**(3), 573–581.

Parnovski, L., and Sobolev, A. 2001b. On the Bethe–Sommerfeld conjecture for the polyharmonic operator. *Duke Math. J.*, **107**(2), 209–238.

Parnovski, L., and Sobolev, A. 2010. Bethe–Sommerfeld conjecture for periodic operators with strong perturbations. *Invent. Math.*, **181**(3), 467–540.

Pastur, L., and Figotin, A. 1992. *Spectra of Random and Almost-Periodic Operators*. Grundlehren der Mathematischen Wissenschaften, vol. 297. Springer-Verlag.

Pavlov, B. 1987. The theory of extensions, and explicitly solvable models. *Uspekhi Mat. Nauk*, **42**(6(258)), 99–131, 247.

Pham, F. 1967. *Introduction à l'Étude Topologique des Singularités de Landau*. Memorial Des Sciences Mathematiques – Fascicule I 164. Gauthier-Villars Éditeur.

Pizzi, G., et al. 2020. Wannier90 as a community code: new features and applications. *J. Phys. Cond. Matt.*, **32**(165902).

Pliś, A. 1960. Non-uniqueness in Cauchy's problem for differential equations of elliptic type. *J. Math. Mech.*, **9**, 557–562.

Popov, V. N., and Skriganov, M. M. 1981. Remark on the structure of the spectrum of a two-dimensional Schrödinger operator with periodic potential. (In Russian.) *Zap. Nauchn. Sem. Leningrad. Otdel. Mat. Inst. Steklov. (LOMI)*, **109**, 131–133, 181, 183–184. English translation, *J. Math. Sci.* **24**, 239–240 (1984).

Post, O. 2003. Periodic manifolds with spectral gaps. *J. Differential Equations*, **187**(1), 23–45.

Prodan, E. 2015. On the generalized Wannier functions. *J. Math. Phys.*, **56**(11), 113511.

Prodan, E., and Schulz-Baldes, H. 2016. *Bulk and Boundary Invariants for Complex Topological Insulators. From K-Theory to Physics*. Mathematical Physics Studies. Springer.

Radin, C., and Simon, B. 1978. Invariant domains for the time-dependent Schrödinger equation. *J. Differential Equations*, **29**(2), 289–296.

Raĭkov, G. D. 1992. Eigenvalue asymptotics for the Schrödinger operator with perturbed periodic potential. *Invent. Math.*, **110**(1), 75–93.

Raikov, G. D. 1998. Asymptotic bounds on the number of the eigenvalues in the gaps of the 2D magnetic Schrödinger operator. *Asymptot. Anal.*, **16**(2), 87–98.

Reed, M., and Simon, B. 1972. *Methods of Modern Mathematical Physics. Vol 1: Functional Analysis*. Academic Press.

Rehr, J. J., and Kohn, W. 1974. Wannier functions in crystals with surfaces. *Phys. Rev. B (3)*, **10**, 448–455.

Reingold, O., Vadhan, S., and Wigderson, A. 2002. Entropy waves, the zig-zag graph product, and new constant-degree expanders. *Ann. Math. (2)*, **155**(1), 157–187.

Rofe-Beketov, F. S. 1963. On the spectrum of non-selfadjoint differential operators with periodic coefficients. *Dokl. Akad. Nauk SSSR*, **152**, 1312–1315.

Rofe-Beketov, F. S. 1964a. A finiteness test for the number of discrete levels which can be introduced into the gaps of the continuous spectrum by perturbations of a periodic potential. *Dokl. Akad. Nauk SSSR*, **156**, 515–518.

Rofe-Beketov, F. S. 1964b. A finiteness test for the number of discrete levels which can be introduced into the gaps of the continuous spectrum by perturbations of a periodic potential. *Dokl. Akad. Nauk SSSR*, **156**, 515–518.

Rofe-Beketov, F. S. 1973. A perturbation of a Hill's operator, that has a first moment and a non-zero integral, introduces a single discrete level into each of the distant spectral lacunae. Pages 158–159, 163 of: *Mathematical Physics and Functional Analysis, No. 4*. (In Russian.) Fiz.-Tehn. Inst. Nizkih Temperatur, Akad. Nauk Ukrain. SSR.

Rofe-Beketov, F. S. 1977. Spectral analysis of the Hill operator and of its perturbations. Pages 144–155 of: *Functional Analysis, No. 9: Harmonic Analysis on Groups*. (In Russian.) Ul′janovsk. Gos. Ped. Inst., Ul′yanovsk.

Rofe-Beketov, F. S. 1979. Generalization of the Prüfer transformation and the discrete spectrum in gaps of the continuous spectrum. Pages 146–153 of: *Spectral Theory of Operators*. Proc. Second All-Union Summer Math. School, Baku, 1975). (In Russian.) "Èlm", Baku.

Rofe-Beketov, F. S. 1984. Spectrum perturbations, the Knezer-type constants and the effective mass of zones-type potentials. Pages 757–766 of: *Constructive Theory of Functions 1984*. Proceedings of the International Conference on Constructive Function Theory, Varna 1984.

Rofe-Beketov, F. S., and Kholkin, A. M. 2005. *Spectral Analysis of Differential Operators*. Interplay between spectral and oscillatory properties, Translated from the Russian by Ognjen Milatovic and revised by the authors, With a foreword by Vladimir A. Marchenko. World Scientific Monograph Series in Mathematics, vol. 7. World Scientific Publishing Co. Pte. Ltd.

Ronkin, L. I. 1974. *Introduction to the Theory of Entire Functions of Several Variables*. Translated from the Russian by Israel Program for Scientific Translations. Translations of Mathematical Monographs, vol. 44. American Mathematical Society.

Rozenblyum, G. V., Solomyak, M. Z., and Shubin, M. A. 1989. Spectral theory of differential operators. Pages 5–248 of: *Current Problems in Mathematics. Fundamental Directions, Vol. 64*. (In Russian.) Itogi Nauki i Tekhniki. Akad. Nauk SSSR, Vsesoyuz. Inst. Nauchn. i Tekhn. Inform.

Rudin, W. 1969. *Function Theory in Polydiscs*. W. A. Benjamin, Inc.

Sakoda, K. 2001. *Optical Properties of Photonic Crystals*. Springer Verlag.

Sánchez-Palencia, E. 1980. *Nonhomogeneous Media and Vibration Theory*. Lecture Notes in Physics, vol. 127. Springer-Verlag.

Scarpellini, B. 1999. *Stability, Instability, and Direct Integrals*. Chapman & Hall/CRC Research Notes in Mathematics, vol. 402. Chapman & Hall/CRC.

Schenker, J., and Aizenman, M. 2000. The creation of spectral gaps by graph decoration. *Lett. Math. Phys.*, **53**(3), 253–262.

Schillinger, M., Mingaleev, S., Hermann, D., and Busch, K. 2005. Highly localized Wannier functions for the efficient modeling of photonic crystal circuits. *SPIE*, **5773**, 324–335.

Shafarevich, I. R. 1994. *Basic Algebraic Geometry. 1*. 2nd ed. Varieties in projective space, Translated from the 1988 Russian edition and with notes by Miles Reid. Springer-Verlag.

Shen, Z. 2001. Absolute continuity of generalized periodic Schrödinger operators. Pages

113–126 of: *Harmonic analysis and Boundary Value Problems (Fayetteville, AR, 2000)*. Contemporary Mathematics, vol. 277. American Mathematical Society.

Shen, Z. 2003. On the Bethe–Sommerfeld conjecture for higher-order elliptic operators. *Math. Ann.*, **326**(1), 19–41.

Shenk, D., and Shubin, M. 1987. Asymptotic expansion of the state density and the spectral function of a Hill operator. *Math. USSR Sbornik*, **56**, 473–490.

Shenk, D., and Shubin, M. A. 1985. Asymptotic expansion of state density and the spectral function of the Hill operator. *Mat. Sb. (N.S.)*, **128(170)**(4), 474–491, 575.

Shenk, D., and Shubin, M. A. 1986. Asymptotic expansion of the spectral function of the Hill operator. *Funktsional. Anal. i Prilozhen.*, **20**(1), 89–90.

Shipman, S. P. 2014. Eigenfunctions of unbounded support for embedded eigenvalues of locally perturbed periodic graph operators. *Commun. Math. Phys.*, **332**(2), 605–626.

Shipman, S. P. 2020. Reducible Fermi surfaces for non-symmetric bilayer quantum-graph operators. *J. Spectr. Theory*, **10**(1), 33–72.

Shnol', È. È. 1957. On the behavior of the eigenfunctions of Schrödinger's equation. *Mat. Sb. (N.S.)*, **42 (84)**(3), 273–286.

Shterenberg, R. 2004. An example of a periodic magnetic Schrödinger operator with a degenerate lower edge of the spectrum. *Algebra i Analiz*, **16**(2), 177–185.

Shterenberg, R. 2005. On the structure of the lower edge of the spectrum of a periodic magnetic Schrödinger operator with small magnetic potential. *Algebra i Analiz*, **17**(5), 232–243.

Shterenberg, R., and Suslina, T. A. 2001. Absolute continuity of the spectrum of the Schrödinger operator with the potential concentrated on a periodic system of hypersurfaces. *Algebra i Analiz*, **13**(5), 197–240. English translation in *St. Petersburg Math. J.* **13**(5), (2002).

Shubin, M. 1979. Spectral theory and the index of elliptic operators with almost-periodic coefficients. *Uspekhi Mat. Nauk*, **34**(2(206)), 95–135.

Shubin, M. 1990. Weak Bloch property and weight estimates for elliptic operators. With an appendix by Shubin and J. Sjöstrand. Pages Exp. No. V, 36 of: *Séminaire sur les Équations aux Dérivées Partielles, 1989–1990*. École Polytechnique, Palaiseau.

Simon, B. 1976. On the genericity of nonvanishing instability intervals in Hill's equation. *Ann. Inst. H. Poincaré Sect. A (N.S.)*, **24**(1), 91–93.

Simon, B. 1979. Phase space analysis of simple scattering systems: extensions of some work of Enss. *Duke Math. J.*, **46**(1), 119–168.

Simon, B. 1996. *Representations of Finite and Compact Groups*. Graduate Studies in Mathematics, vol. 10. American Mathematical Society.

Simon, B. 2005. *Trace Ideals and their Applications*. 2nd ed. Mathematical Surveys and Monographs, vol. 120. American Mathematical Society.

Simon, B. 2011. *Szego's Theorem and its Descendants, Spectral Theory for L^2 Perturbations of Orthogonal Polynomials*. Princeton University Press.

Sivaji Ganesh, S., and Tewary, V. 2020. Generic simplicity of spectral edges and applications to homogenization. *Asymptot. Anal.*, **116**(3-4), 219–248.

Sjöstrand, J. 1991. Microlocal analysis for the periodic magnetic Schrödinger equation and related questions. Pages 237–332 of: *Microlocal Analysis and Applications (Montecatini Terme, 1989)*. Lecture Notes in Mathematics, vol. 1495. Springer.

Skriganov, M. 1985a. *Geometric and Arithmetic Methods in the Spectral Theory of Multidimensional Periodic Operators*. Vol. 171. Mat. Inst. Steklov.

Skriganov, M. 1985b. The spectrum band structure of the three-dimensional Schrödinger operator with periodic potential. *Invent. Math.*, **80**(1), 107–121.

Skriganov, M., and Sobolev, A. 2005. Asymptotic estimates for spectral bands of periodic Schrödinger operators. *Algebra i Analiz*, **17**(1), 276–288.

Slater, J. (ed). 1972. *Symmetry and Energy Bands in Crystals*. Dover.

Smyshlyaev, V. P., and Kuchment, P. 2007. Slowing down and transmission of waves in high contrast periodic media via "non-classical" homogenization. Pages 200–202 of: *Proc. 8th Int. Conf. on Math. & Numerical Aspects of Wave Propagation, Univ. of Reading*.

Sobolev, A. 1999. Absolute continuity of the periodic magnetic Schrödinger operator. *Invent. Math.*, **137**(1), 85–112.

Sobolev, A. 2003. High energy asymptotics of the density of states for certain periodic pseudo-differential operators in dimension one. Pages 171–184 of: *Waves in Periodic and Random Media (South Hadley, MA, 2002)*. Contemporary Mathematics, vol. 339. American Mathematical Society.

Sobolev, A. 2006. Asymptotics of the integrated density of states for periodic elliptic pseudo-differential operators in dimension one. *Rev. Mat. Iberoam.*, **22**(1), 55–92.

Sobolev, A. 2007. Recent results on the Bethe–Sommerfeld conjecture. Pages 383–398 of: *Spectral Theory and Mathematical Physics: A Festschrift in Honor of Barry Simon's 60th Birthday*. Proceedings of Symposia in Pure Mathematics, vol. 76. American Mathematical Society.

Sobolev, A. V. 1991. Weyl asymptotics for the discrete spectrum of the perturbed Hill operator. Pages 159–178 of: *Estimates and Asymptotics for Discrete Spectra of Integral and Differential Equations (Leningrad, 1989–90)*. Advances in Soviet Mathematics, vol. 7. American Mathematical Society.

Sobolev, A. V., and Walthoe, J. 2002. Absolute continuity in periodic waveguides. *Proc. London Math. Soc. (3)*, **85**(3), 717–741.

Soluyanov, A. A., and Vanderbilt, D. 2011. Wannier representation of Z2 topological insulators. *Phys. Rev. B*, **83**(3), 035108.

Souza, I., Marzari, N., and Vanderbilt, D. 2001. Maximally localized Wannier functions for entangled energy bands. *Phys. Rev. B*, **65**(3), 035109.

Stein, E. M., and Weiss, G. 1971. *Introduction to Fourier analysis on Euclidean spaces*. Princeton Mathematical Series, No. 32. Princeton University Press.

Stigler, S. 1980. Stigler's law of eponymy. *Trans. NY Acad. Sci.*, **39**(1), 147–157.

Stollmann, P. 2001. *Caught by Disorder. Bound States in Random Media*. Progress in Mathematical Physics, vol. 20. Birkhäuser Boston, Inc.

Strang, G. 1980. *Linear Algebra and Its Applications*. 2nd ed. Academic Press

Sunada, T. 1985. Riemannian coverings and isospectral manifolds. *Ann. Math. (2)*, **121**(1), 169–186.

Sunada, T. 1988. Fundamental groups and Laplacians. Pages 248–277 of: *Geometry and Analysis on Manifolds (Katata/Kyoto, 1987)*. Lecture Notes in Mathematics, vol. 1339. Springer.

Sunada, T. 1990. A periodic Schrödinger operator on an abelian cover. *J. Fac. Sci. Univ. Tokyo Sect. IA Math.*, **37**(3), 575–583.

Sunada, T. 1992. Group C^*-algebras and the spectrum of a periodic Schrödinger operator on a manifold. *Canad. J. Math.*, **44**(1), 180–193.

Sunada, T. 2008. Discrete geometric analysis. Pages 51–83 of: *Analysis on Graphs and its Applications*. Proceedings of Symposia in Pure Mathematics, vol. 77. American Mathematical Society.

Suslina, T. A. 2001. On the absence of eigenvalues of a periodic matrix Schrödinger operator in a layer. *Russ. J. Math. Phys.*, **8**(4), 463–486.

Suslina, T. A. 2002. Absolute continuity of the spectrum of the periodic Maxwell operator in a layer. (In Russian.) *Zap. Nauchn. Sem. S.-Peterburg. Otdel. Mat. Inst. Steklov. (POMI)*, **288**, 232–255. English translation: *J. Math. Sci. (NY)*, **123**(6) (2004), 5654–4667.

Suslina, T. A. 2003a. The discrete spectrum of a two-dimensional second-order periodic elliptic operator perturbed by a decaying potential. II. Inner gaps. *Algebra i Analiz*, **15**(2), 128–189.

Suslina, T. A. 2003b. On discrete spectrum in the gaps of a two-dimensional periodic elliptic operator perturbed by a decaying potential. Pages 185–199 of: *Waves in Periodic and Random Media (South Hadley, MA, 2002)*. Contemporary Mathematics, vol. 339. American Mathematical Society.

Tao, T. 2011. The limiting absorption principle. `https://terrytao.wordpress.com/2011/04/21/the-limiting-absorption-principle`.

Taylor, M. E. 2011. *Partial Differential Equations I. Basic Theory*. 2nd ed. Applied Mathematical Sciences, vol. 115. Springer.

Thomas, L. 1973. Time dependent approach to scattering from impurities in a crystal. *Commun. Math. Phys.*, **33**, 335–343.

Thouless, D. J. 1984. Wannier functions for magnetic sub-bands. *J. Phys. C: Solid State Phys.*, **17**, L325–7.

Thouless, D. J. 1998. *Topological Quantum Numbers in Nonrelativistic Physics*. World Scientific.

Tikhomirov, M., and Filonov, N. 2004. Absolute continuity of an "even" periodic Schrödinger operator with nonsmooth coefficients. *Algebra i Analiz*, **16**(3), 201–210.

Titchmarsh, E. C. 1958. *Eigenfunction Expansions Associated with Second-Order Differential Equations. Vol. 2*. Oxford University Press.

Vaĭnberg, B. R. 1966. Principles of radiation, limiting absorption and limiting amplitude in the general theory of partial differential equations. *Uspehi Mat. Nauk*, **21**(3 (129)), 115–194.

Van Hove, L. 1953. The occurrence of singularities in the elastic frequency distribution of a crystal. *Phys. Rev.*, **89**(Mar), 1189–1193.

Veliev, O. A. 1986. Spectral expansion of nonselfadjoint differential operators with periodic coefficients. *Differentsial'nye Uravneniya*, **22**(12), 2052–2059.

Veliev, O. A. 1988. The spectrum of multidimensional periodic operators. *Teor. Funktsiĭ Funktsional. Anal. i Prilozhen.*, **49**, 17–34, 123.

Veliev, O. A. 2007. Perturbation theory for the periodic multidimensional Schrödinger operator and the Bethe–Sommerfeld conjecture. *Int. J. Contemp. Math. Sci.*, **2**(1-4), 19–87.

Veliev, O. A. 2009. On the constructive determination of the periodic potentials from the Bloch eigenvalues. *J. Phys. A*, **42**(37), 375201, 19.

Veliev, O. A. 2011. An algorithm for finding the periodic potential of the three-dimensional Schrödinger operator from spectral invariants. *J. Phys. A*, **44**(15), 155202, 25.

Veliev, O. A. 2024. *Multidimensional Periodic Schrödinger Operator – Perturbation Theories for High Energy Regions and Their Applications*. 3rd ed. Springer Tracts in Modern Physics, vol. 291. Springer.

Von Neumann, J., and Wigner, E. P. 1929. Über Merkwürdige Diskrete Eigenwerte. *Z. Phys.*, **30**, 465–467.

Vorobets, M. 2011. On the Bethe–Sommerfeld conjecture for certain periodic Maxwell operators. *J. Math. Anal. Appl.*, **377**(1), 370–383.

Wallace, P. R. 1947. The band theory of graphite. *Phys. Rev.*, **71**, 622.

Wannier, G. H. 1937. The structure of electronic excitation levels in insulating crystals. *Phys. Rev.*, **52**, 191–197.

Weyl, H. 1909. Über beschrankte quadratische Formen, deren Differenz vollstetig ist. *Rend. Circ. Mat. Palermo*, **27**, 373–392.

Weyl, H. 1989. *Symmetry*. Princeton Science Library. Princeton University Press. Reprint of the 1952 original.

Wu, W., et al. 2024. The discovery of three-dimensional Van Hove singularity. *Nat. Commun.*, **15**, 2313.

Yablonovitch, E. 1987. Inhibited spontaneous emission in solid-state physics and electronics. *Phys. Rev. Lett.*, **58**(20), 2059–2062.

Yakubovich, V. A., and Starzhinskii, V. M. 1975. *Linear differential equations with periodic coefficients. V. 1, 2*. Halsted Press [John Wiley & Sons] Translated from Russian by D. Louvish.

Yariv, A., Xu, Yong, Lee, R., and Scherer, A. 1999. Coupled-resonator optical waveguide: a proposal and analysis. *Opt. Lett.*, **24**, 711–713.

Yates, J., Wang, X., Vanderbilt, D., and Souza, I. 2007. Spectral and Fermi surface properties from Wannier interpolation. *Phys. Rev. B*, **75**, 195121.

Yudovich, V. I. 1989. *The linearization method in hydrodynamical stability theory*. Translations of Mathematical Monographs, vol. 74. American Mathematical Society, Providence, RI. Translated from the Russian by J. R. Schulenberger.

Zaidenberg, M., Krein, S. G., Kuchment, P., and Pankov, A. 1975. Banach bundles and linear operators. *Uspehi Mat. Nauk*, **30**(5(185)), 101–157. English translation: Russian Math. Surveys 30 (1975), no. 5, 115–175.

Zelditch, S. 1990. On the generic spectrum of a Riemannian cover. *Ann. Inst. Fourier (Grenoble)*, **40**(2), 407–442.

Želobenko, D. P. 1973. *Compact Lie groups and Their Representations*. Translated from the Russian by Israel Program for Scientific Translations. Translations of Mathematical Monographs, vol. 40. American Mathematical Society.

Želudev, V. A. 1967. The eigenvalues of a perturbed Schrödinger operator with periodic potential. Pages 108–123 of: *Problems of Mathematical Physics, No. 2, Spectral Theory, Diffraction Problems*. (In Russian.) Izdat. Leningrad. Univ.

Želudev, V. A. 1968. The perturbation of the spectrum of the Schrödinger operator with a complex-valued periodic potential. Pages 31–48 of: *Problems of Mathematical Physics, No. 3: Spectral Theory*. (In Russian.) Izdat. Leningrad. Univ.

Želudev, V. A. 1970. The perturbation of the spectrum of the one-dimensional selfadjoint Schrödinger operator with periodic potential (errata.) Pages 61–82 of: *Problems of Mathematical Physics, No. 4: Spectral Theory. Wave Processes*. (In Russian.) Izdat. Leningrad. Univ.

Zhelobenko, D. 2006. *Principal Structures and Methods of Representation Theory*. Translated from the 2004 Russian original by Alex Martsinkovsky. Translations of Mathematical Monographs, vol. 228. American Mathematical Society.

Zhikov, V. V. 2004. Gaps in the spectrum of some elliptic operators in divergent form with periodic coefficients. *Algebra i Analiz*, **16**(5), 34–58.

Ziman, J. M. 1962. *Electrons in Metals: A Short Guide to the Fermi Surface*. Taylor & Francis.

Ziman, J. M. 1964. *Principles of the Theory of Solids*. Cambridge University Press.

Zimbovskaya, N. A. 2001. *Local Geometry of the Fermi Surface: And High-Frequency Phenomena in Metals*. Springer Verlag.

Index

For EU product safety concerns, contact us at Calle de José Abascal, 56–1°, 28003 Madrid, Spain or eugpsr@cambridge.org.

www.ingramcontent.com/pod-product-compliance
Lightning Source LLC
LaVergne TN
LVHW020340240926
848761LV00004B/15
9781009282420